M000268035

Measuring the Performance of Biometric Systems

Measurement	Abbreviation	Description
False Accept Rate	FAR	The biometric system incorrectly declares a successful match for an invalid subject.
False Reject Rate	FRR	The biometric system incorrectly declares a mismatch for a valid subject.
Equal Error Rate	ERR	The point at which the FAR and the FRR are equal.
Failure to Enroll	FER	The biometric system cannot collect a sample when a new subject enrolls.
Failure to Capture	FTC	The biometric system fails to recognize a biometric when one is presented.

Biometrics Basics

Biometrics are used for both *authentication* (to show that a person matches a presented ID) and identification (to identify someone using only a presented biometric sample).

Biometrics can be classified into two basic types: *physiological*, which uses physical characteristics of some part of the body or *behavioral*, which uses some aspect of learned behavior.

The most accurate biometrics available usually focus on one or more details not apparent to the human eye, but easily discernable to a computer. For this reason, iris recognition is very accurate and facial recognition is less so.

Biometric systems are particularly good at providing *non-repudiation* — in effect, proving that the authenticated person really performed an action, and not an impostor.

Basic Biometric Type	Depends on...	Effectiveness	Includes...
Behavioral	Users performing well-known tasks (such as writing or walking) in very similar ways every time.	The more a behavioral biometric is used, the more accurate it will be as an authentication or identification tool.	Signature, voice, keystroke, gait
Physiological	Detailed information about parts of the body to uniquely identify or authenticate a person.	The best physiological biometrics are those that change very little over time and are protected from damage, such as those based on the iris or hand veins.	Fingerprint, hand scan, iris scan, retina scan, facial scan

Biometrics For Dummies®

Biometrics Acceptance, Privacy, and Law

- People are most comfortable with biometric collection systems that are the least intrusive. Retinal scanners and electronic noses are a bit too intrusive; iris imaging and touch-free hand-vein scanners are more comfortable.

- Commonly, the information stored by biometric systems could not be used to recreate an image, but re-creating a fingerprint (or other biometric) from stolen data is a common fear.

- Touch-based biometric sensors (such as fingerprint, palm print, and hand geometry) can be disease vectors unless sanitary precautions are taken. Oddly, they are no less sanitary than doorknobs, but doorknobs are better accepted.

- Stolen biometric data can typically be used only if the attacker can inject that data directly into the information flow of an authentication transaction via the network or wires from the sensor.

- Some kinds of biometric data (such as fingerprints, facial images, and gait characteristics) are exposed to attackers' attempts to collect them from such sources as drinking glasses, camera phones, and video cameras.

- Some kinds of biometrics, such as those obtained from the retina, iris, and hand veins, can potentially reveal medical data to the organization (in particular, health changes when the system detects changes in these readings).

- In the United States, few laws actually offer direct protection for the privacy of biometric information — and companies that collect such information are typically not under any obligation to disclose the loss of it (usually a result of hacking or theft).

- In the European Union, privacy laws protect the collection and subsequent use of personal information, including biometric information.

For Dummies: Bestselling Book Series for Beginners

Biometrics For Dummies®

Published by
Wiley Publishing, Inc.
111 River Street
Hoboken, NJ 07030-5774

www.wiley.com

For general information on our other products and services, please contact our Customer Care Department within the U.S. at 800-762-2974, outside the U.S. at 317-572-3993, or fax 317-572-4002.

For technical support, please visit www.wiley.com/techsupport.

Wiley also publishes its books in a variety of electronic formats. Some content that appears in print may not be available in electronic books.

Library of Congress Control Number: 2008930830

ISBN: 978-0-470-29288-4

Manufactured in the United States of America

10 9 8 7 6 5 4 3 2 1

WILEY

Biometrics

FOR

DUMMIES®

**by Peter Gregory, CISA, CISSP
and Michael A. Simon**

Wiley Publishing, Inc.

About the Authors

Peter Gregory, CISA, CISSP, is the author of several books including *IT Disaster Recovery Planning For Dummies, Blocking Spam & Spyware For Dummies* (with Mike Simon) and *CISSP For Dummies.*

Peter is the security and risk manager at a financial management software company located in Redmond, Washington. Prior to this, he held tactical and strategic security positions in large wireless telecommunications organizations. He has also held development and operations positions in casino gaming-management systems, banking, government, nonprofit organizations, and academia since the late 1970s. He is a member of the Board of Advisors and an occasional lecturer for the NSA-certified University of Washington Certificate Program in Information Assurance & Cybersecurity.

Peter can be found at www.peterhgregory.com.

Michael A. Simon is the author of *The Internet Starter Kit for Windows* (with Adam Engst and Corwin S. Low) and *Blocking Spam & Spyware For Dummies* (with Peter Gregory).

Mike has been working in computer security and policy development since 1985, working at the time for the University of Idaho, a regional pioneer in computer security and one of the first NSA Centers of Excellence in Information Assurance Education.

Currently, Mike is an adjunct faculty member for the University of Washington, and occasionally lectures at Seattle University, University of Idaho, and several civic organizations on the subject of information assurance and computer security. He sits on the advisory board for the Information Assurance certificate program for the University of Washington, the technical advisory board for Goldfish Holdings, Inc., the Advisory Board for the Computer Science Department at the University of Idaho, and on the Founders Board for the Information School at the University of Washington.

Dedication

To Becky and Shannon. — Peter Gregory

To my teachers: past, present, and future. — Mike Simon

Authors' Acknowledgments

Peter Gregory would like to thank Carole McClendon, his literary agent, and Tiffany Ma and Amy Fandrei, Acquisition Editors at Wiley, for their support of this project. Thank you to Nicole Sholly, Project Editor at Wiley, for your help organizing our work, and to Barry Childs-Helton and John Chirillo for copy and technical editing, respectively. Thank you, Mike, I always enjoy working with you on collaborative projects.

Mike Simon would like to thank Paul Donion for dealing with a business partner with deadlines. Thanks to Erin Klunder and Ray Pompon for answering random biometrics questions about law enforcement and finance (respectively). Much thanks to Al Gidari and Joseph Cutler of Perkins Coie, LLP for the use of the table of State Data Breach laws in Chapter 3. Thanks, Peter, for making me look good (again).

Publisher's Acknowledgments

We're proud of this book; please send us your comments through our online registration form located at www.dummies.com/register/.

Some of the people who helped bring this book to market include the following:

Acquisitions and Editorial

Project Editor: Nicole Sholly

Acquisitions Editor: Amy Fandrei

Senior Copy Editor: Barry Childs-Helton

Technical Editor: John Chirillo

Editorial Manager: Kevin Kirschner

Editorial Assistant: Amanda Foxworth

Senior Editorial Assistant: Cherie Case

Cartoons: Rich Tennant
(www.the5thwave.com)

Composition Services

Senior Project Coordinator: Kristie Rees

Layout and Graphics: Reuben W. Davis, Joyce Haughey, Melissa K. Jester, Abby Westcott, Christine Williams

Proofreaders: Dwight Ramsey, Nancy L. Reinhardt

Indexer: Claudia Bourbeau

Publishing and Editorial for Technology Dummies

　　Richard Swadley, Vice President and Executive Group Publisher

　　Andy Cummings, Vice President and Publisher

　　Mary Bednarek, Executive Acquisitions Director

　　Mary C. Corder, Editorial Director

Publishing for Consumer Dummies

　　Diane Graves Steele, Vice President and Publisher

　　Joyce Pepple, Acquisitions Director

Composition Services

　　Gerry Fahey, Vice President of Production Services

　　Debbie Stailey, Director of Composition Services

Contents at a Glance

Table of Contents

Introduction

. . . .

*B*iometric technology is a HOT topic. The most recent (ISC)² Global Workforce Survey of over 7,500 security professionals from around the world have cited biometrics as the number one security project for organizations in North America, and the number two security project in the world overall.

Most companies aren't using any form of biometrics yet, but many are looking into biometrics now. Security regulation, negative publicity about security breaches, and the desire to avoid being the *next* security-breach scandal are prompting many organizations to take a first serious look at biometrics.

About This Book

Our objective for this book is simple: to help you become familiar with biometrics in a hurry by giving you all the information you need to be able to start looking at biometric solutions. *Biometrics For Dummies* will help you to be a smart shopper and know what you are looking for — and what to avoid.

This book isn't the encyclopedia of biometrics. You don't need that right now. Given the broad base of information you need, this book is the best way to jump-start your own knowledge, which will help you and your company become critical, informed customers. You'll learn what biometrics are all about, all the general categories of biometrics (plus a few of the oddities), and everything you need to know to get started with an evaluation and eventual implementation of biometrics.

As a technology professional or a business manager, you'll find that biometrics are a bit tricky, especially with respect to privacy and legal issues, but not really all that hard to learn. Learning about biometrics isn't any more difficult than learning any new hobby or skill. And as you proceed down the road to evaluating, selecting, and using biometrics, you and your organization will be far better off for all the insight this book imparts. You'll probably get some kudos for helping your organization successfully embrace and use biometrics. We have confidence in you!

Foolish Assumptions

We make two broad presumptions when we wrote this book:

- ✔ **You understand your business.** You know what your business's products and services are, and how it develops, sells, and supports them. You don't need to understand every minute detail, but just have a good overall view of what your business is *in* the business of doing.

- ✔ **You understand technology.** We aren't saying you're a rocket surgeonist (say what?), but we figure you know the basics of how computers, networks, and applications work — and what's going on when you type in your user ID and password or swipe your card when you enter the building.

Conventions Used in This Book

This isn't a book about computer programming or acupuncture, so we've spared you a lot of tricky diagrams and lines of computer code. We've written *Biometrics For Dummies* in plain English, and that's about all you need to know! (Okay, we have used one convention: We've italicized jargon-y terms as a heads-up to you that the jargon-y term's definition — as it pertains to biometrics — is nearby.)

What You Don't Have to Read

If your organization already has a biometrics system and you need to know more about how it works, then you can probably skip the chapters on selection and implementation, unless you'd like some insight into how the decision-making process probably went in your company.

How This Book Is Organized

This book is organized in four parts. Although the chapters don't necessarily have to be read in order, they're organized according to the somewhat logical progression that an organization would follow as it explores its security issues and proceeds down the road to Biometric City.

Part 1: Getting Started with Biometrics

In Part I, we first introduce biometrics and give you some sweeping overviews on how biometrics help to protect organization assets. We also discuss the impact that biometrics are having on data security and privacy laws, as well as professional ethics. The chapters in this part give you a solid business background on what biometric technology is all about — and the impact it's having on business and society.

Part 11: Types of Biometrics

Part II explains all the types of biometrics that are in use today, as well as some that are still up and coming. The chapters in this part are organized by the type of biometrics: fingerprint and hand, signature, ocular and facial, and other types.

If you've already settled on a type of biometrics that will work in your organization, we still suggest you read all the chapters in this section. Even if your educated guess was right, learning more about the other types of biometrics will give you keen insight and will help open your mind to consider alternatives.

Part 111: Implementing and Using Biometrics

Selecting biometrics is only half the fun; once you've made an educated and informed decision on a biometrics solution, you've got to make it work. We've poured our decades of technology-implementation and management expertise into the chapters in this part to help you avoid pitfalls we discovered long ago. Digging yourself into a hole is no fun, especially when others are looking. But rather than giving you a shovel to make the hole, or even the ladder to climb out of it, we provide a map to help you sidestep it altogether.

In this part, we also discuss how to protect your biometrics system from harm. Biometric technology itself needs to be protected, so it can perform properly and protect your company's assets. Finally, we discuss where we think biometric technology is going in the future. This will help jump-start you on some of the conventional — and not-so-conventional — wisdom in the biometrics industry and practices.

Part IV: The Part of Tens

No *For Dummies* book is complete without a Part of Tens. And the Part of Tens chapters in *Biometrics For Dummies* are really special. You won't find lists of Web sites here; instead, you'll marvel, be entertained, and wind up informed. And you'll probably go back and give the Part of Tens a second look in the other *For Dummies* books on your bookshelf. It's really that good!

Part V: Appendixes

This part contains a great consumer guide comparing all the biometric technologies discussed in this book and an appendix that covers physical security for IT pros who have spent most of their careers paying attention to computers. Just before the index is a short listing of biometric and information security terms. We've included this glossary in the book so you can refer to it often and easily whenever some wily biometric term escapes you.

Icons Used in This Book

What's a *For Dummies* book without icons pointing you in the direction of really great information that's sure to help you along your way? Icons are great visual cues to handy information; here's a brief description of each icon we use in this book:

The Tip icon points out helpful information that's likely to make your job easier.

This icon marks a generally interesting and useful fact that you may want to remember for later use, because it's likely to crop up again.

The Warning icon highlights lurking danger. With this icon, we're telling you to pay attention and proceed with caution.

Where to Go from Here

If you need to learn about the various types of biometrics, turn to Part II. If your organization has already chosen a solution and you need to help implement it, go to Chapter 9. If you're wondering about the impact of biometrics on law and society, read Chapter 3. If you want to look over our shoulder into the biometrics crystal ball, go to Chapter 11. You can dog-ear the glossary in case you want to learn the language of biometrics.

Write to Us!

We're all too familiar with the fact that technology and business marches on, and we won't be left in the dust. If you're embarking on your own biometrics project and have questions, or have your own wisdom to share, look us up on the Internet or write to us here:

peterhgregory@yahoo.com

mikeasimon@gmail.com

Part I

Getting Started with Biometrics

The 5th Wave   By Rich Tennant

"We take network security here very seriously."

In this part . . .

So you're getting started with biometrics, eh? Maybe your organization is implementing a biometric security solution, and perhaps you're even in charge of it! You've come to the right place: In this section, you get a quick overview of all things biometric, as well as in-depth discussions of how biometrics are used to protect assets — and a look at laws that are related to biometrics.

Chapter 1

Understanding Biometrics

*H*ere's our "nickel tour" of biometrics — well, okay, that'd be a dollar or two in today's money — back in the day, a nickel tour meant you got a pretty good overall look at something in a short time. This chapter is like that. If you're going into a meeting about biometrics in thirty minutes, and you don't want to appear clueless about it, this chapter will show you all the basics you need. But if you're standing in the bookstore looking for more of a clue, you're much better off buying the book and taking it home where you can drill deeper into the topics you're really interested in.

What Biometrics Are and Who's Using Them

The term *biometrics* comes from the ancient Greek *bios* = "life" and *metron* = "measure." Biometrics refers to the entire class of technologies and techniques to uniquely identify humans. Though biometric technology has various uses, its primary purpose is to provide a more secure alternative to the traditional access-control systems used to protect personal or corporate assets. Many of the problems that biometrics help to solve are the weaknesses found in present access-control systems — specifically these:

✔ **Weak passwords:** Computer users are notoriously apt to use poor, easily guessed passwords, resulting in break-ins where intruders can guess another user's credentials and gain unauthorized access to a computer system. This could lead to a security breach where personal or business secrets are stolen by an outsider. If your password is currently *password, 123456, abc123, letmein,* or *qwerty,* please stop reading this book long enough to change it — to the first character of each word in your favorite passage from *Moby Dick*. We'll wait.

✔ **Shared credentials:** In both small and large organizations, we often hear of cases like this: A computer user shares his or her password with a colleague who requires access — even though, in most organizations (and in many security-related laws and regulations), this is forbidden by policy. People by nature are willing to help a colleague in need, though, even if it means violating policy to achieve a greater purpose.

✔ **Lost key cards:** Many times in our careers we have both found lost key cards in parking lots and other places. Often they have the name of the organization on them, so it's like finding a key with an address on it, permitting the person who found it a free after-hours tour of some American corporation.

Biometrics can solve all these problems by requiring an additional credential — something associated with the person's own body — before granting access to a building, computer room, or computer system. An access-control system that utilizes biometrics will include an electronic device that measures some specific aspect of a person's body or behavior that positively identifies that person. The device might be a fingerprint reader, a digital camera to get a good look at an iris, or a signature pad. (We discuss all the common types of biometrics in the next section.)

Biometric technology as a means of protecting assets has been around for quite a while in some fields. Military, intelligence, and law enforcement organizations have been using biometrics to enhance physical and logical access controls for decades.

But in the past several years, there has been an uptick in the use of biometrics to protect high-value assets. Internet data centers (the kind that lease rack space and cage space to companies that prefer not to build their own fortresses) often use biometrics for admitting personnel to the data-center floor. Fingerprint-biometric devices are showing up everywhere — even built in to laptops, PDAs, and USB drives. Facial recognition is available on a few laptop models. And for protecting businesses and residences, fingerprint-biometric door-lock sets are available at your favorite big-box home-improvement center (though most of these have key-based bypass systems, reducing the *actual* security you get to the level of a key-based system).

We've also seen a grocery-store chain here in Seattle experiment with using fingerprint scanners for checkout-line payment. Walt Disney World in Orlando, Florida uses fingerprint readers for customers who purchase multi-day passes, to ensure that those who reenter the facility on subsequent days are the same people who purchased the tickets on the first day. Everyone who attended Super Bowl XXXV had their faces compared to the faces of known criminals, using biometrics. Anyone entering the United States since September 30, 2004, has submitted prints of both index fingers — and in December 2008 that will extend to all prints from both hands.

Types of Biometrics

Although there are close to a dozen more-or-less effective ways to use biometrics to identify someone, they all fall into two classes (see Figure 1-1): physiological and behavioral.

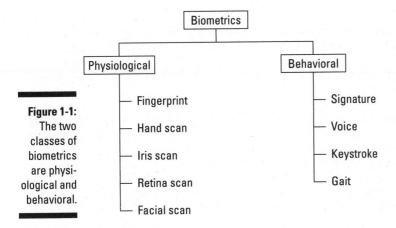

Figure 1-1:
The two classes of biometrics are physiological and behavioral.

Physiological

Physiological biometrics measure a specific part of the structure or shape of a portion of a subject's body. The types of physiological biometrics include:

- **Fingerprint:** Officially established as a means of uniquely identifying people since around 1900, fingerprints are easily registered and measured — and devices for doing so are small and inexpensive. You can find them built in to laptop computers, PDAs, USB drives, door locks, and even credit cards.

- **Hand scan:** The geometry of an entire human hand is quite unique, almost as much as fingerprints themselves. Usually a hand scan does not measure the fingerprint-like patterns in the fingers and palms, but instead relies on the lengths and angles of fingers, the geometry of the entire collection of 27 bones, plus muscles, ligaments, and other tissues.

- **Hand veins:** If you shine a bright light through your hand, you can see an interesting pattern of veins — and also the bones and other elements in your hand.

- **Iris scan:** The human iris is the set of muscles that control the size of the pupil — that little hole in the middle of your eye. The human iris, when viewed up close, is the complex collection of tiny muscles that are stained various colors of brown, gray, blue, and green. When we say that someone has blue, green, or brown "eyes," the color we're referring to is the color of the iris.

- **Retina scan:** The retina is the surface at the rear of the interior of the eye. It's not normally seen except when (say) a doctor shines a bright light through the pupil just right. But it does show up when you have a photo with "red eye" — that's the reflection of the retina. Red eye is not sufficient to identify someone; instead, it is necessary for a person to get their eye close up to a little camera that can see inside the eye.

- **Face recognition:** We recognize faces almost from birth, although *how* we recognize them is better understood now, enough that we can teach computers how to do it under certain conditions. Some laptop computers use facial recognition as a form of authentication before a subject can access the computer.

The characteristic in common with physiological biometrics is that they're more-or-less static measurements of a specific part of your body. You might have to swipe your finger, place your hand, or look at the red dot, but the biometric equipment does the rest. Just hold still . . . there, *got it.*

Physiological biometrics are discussed in detail in Part II. There you can also read about some of the unusual biometrics that may be used someday.

Behavioral

Behavioral biometrics are more concerned with how you *do* something, rather than just a static measurement of a specific body part. Some of the behavioral biometrics in use include these:

- **Handwriting:** Everyone's handwritten signature is different, probably uniquely so. Biometric systems measure signatures in a number of different ways:

  - *Static image.* This is the oldest type of handwriting recognition where we compare a stored signature image with a new sample to see if they match. Arguably, with practice, the *image* of someone's signature can be forged, although it's extremely unlikely that the forger will create the signature the same way that the original person does, which leads to the next two forms of handwriting biometrics:

  - *Signature dynamics.* Here we're measuring either (a) the motion of the stylus or pen or (b) the dynamics of how the signature image itself is created.

  - *Stylus pressure.* We can also measure the dynamics of the downward force of the stylus on the writing surface while the signature is being made.

✔ **Keystroke dynamics:** The rhythm of someone's typing (or *keyboarding* as we tend to call it these days) is as unique as someone's signature. The precise timing of individual keystrokes is a product of the geometry of the hand, the tone of the muscles in the hands and forearms, as well as the brain's ability to send out the right signals at the right time that result in (say) Peter typing this sentence. And one nice thing about keystroke dynamics biometrics is that it's entirely passive — a software program can continually measure keystrokes — and can, in many cases, sense whether someone walked away and someone else sat down and continued typing.

✔ **Voice recognition:** As with typing, the *sound* of someone's voice is the product of physical characteristics (specifically, the construction of larynx and related passageways); the brain is in control of the *linguistics* of one's voice. Some biometric systems will have the subject speak his or her name or password; better ones might have the subject read a unique phrase such as "Liberty requires virtue and mettle."

✔ **Gait:** The way a person walks forms a unique pattern that can be captured for biometric purposes. As with facial recognition, it's sometimes easy to recognize particular people at a distance by the way they walk.

There are stranger and more interesting behavioral biometrics that we discuss more fully in Part II.

How Biometric Systems Work

Biometric systems work through enrolling users by measuring and storing their particular biometric, and then later comparing the stored biometric data with data from unverified subjects to determine whether they should be allowed to access a system or location. Take a look at the entire process in more detail:

1. **Enrollment.**

   Before a user can begin using a biometric system, he or she must complete an enrollment process. Depending upon the biometric technology in use, the user might do this on her own, or there may be a facilitator to help. The user provides other information such as her user ID or name, and then provides initial biometric data, which could consist of (for example) swiping fingers over a fingerprint reader (for fingerprint biometrics), looking into a digital camera lens (for iris biometrics), or repeating some words or phrases (for voice biometrics). Usually the biometric system will request several samples so that the system can determine an average and deviation.

2. **Usage.**

   When the user wishes to access a system or building guarded with biometrics, the user authenticates according to procedure, which could mean swiping a finger over a biometric fingerprint reader, placing a hand over a hand scanner, or signing his name. However it's done, the biometric system will compare the sample with data stored at enrollment time, and make a go/no-go decision on whether the biometric data matches or not. If there is a match, the user is given access; if not, he is denied access and given another try.

3. **Update.**

   For the type of biometrics that change slowly over time (such as handwriting or facial recognition), the biometric system may need to update the data that was originally submitted at enrollment. The biometric system may perform this update with each subsequent measurement (thereby increasing the number of samples, with emphasis on the newer ones), or it may utilize a separate update process.

Biometric systems are generally pretty easy to use. In most cases, even enrollment takes only a minute or two, and everyday usage takes only a few seconds. Indeed, regular use may take less time than the old way of gaining access to a computer or building. This is why we usually consider biometrics a break-even in terms of the time required to use the system compared to the former way in which someone had to identify themselves.

Characteristics of Biometric Systems

Every type of biometric measurement can be classified with a number of characteristics that should be considered in a selection process. Being familiar with these characteristics will help you to better understand how to think objectively about each type. Sure, some of the available biometric technologies are cool, but it's no longer the 1990s — we have to make *rational* decisions about purchasing and using technology. Anyway, the characteristics we're talking about are

- **Universality:** This refers to whether each person has the characteristic being measured. For instance, nearly everyone in your organization will have at least one finger for fingerprint biometrics, but gait-based biometrics may be more difficult if you have any wheelchair-bound staff members.

- **Uniqueness:** How well the particular biometric distinguishes people. DNA is the best, and fingerprints and iris scans are pretty good too.

- **Permanence:** A good biometric system should measure something that changes slowly (if at all) over time. DNA and fingerprints are very good over the long term; handwriting and voice change somewhat from decade to decade.

✔ **Collectability:** This refers to how easily the biometric can be measured. DNA scores very low (it isn't easy to collect); fingerprint and palm-scan biometrics rate quite high. Gait requires a person to walk over a distance, which would be hard to do while sitting at a workstation. Retina scan requires the subject get really close to a digital camera.

✔ **Performance:** This refers to the overall technology burden: how much equipment, time, and calculation go into performing a comparison. The fingerprint method fares very well; fingerprint readers are small, compact, and accurate. DNA biometrics tend to be costly, slow, and labor-intensive.

✔ **Accuracy:** How well does a biometric system distinguish between subjects, and what are the false acceptance and false rejection rates?

✔ **Acceptability:** Will users be willing to use the biometric technology? DNA will score low because of privacy reasons. Retina scans will score low because some people will be uncomfortable putting their eye really close to something that seems intrusive. Similarly, people won't mind swiping a finger across a surface-type fingerprint scanner or getting an iris photographed from a few feet away, but some are squeamish about sticking their fingers into a device (too many "B" movies).

✔ **Circumvention:** This refers to how easily a forgery can be made that will fool the biometric system (early fingerprint devices, for example, could be fooled with "gummy fingers"). *Proof of life* testing — a feature that determines whether a sample comes from a *living* body part — is incorporated into many biometric systems so digital images of body parts are less likely to fool the system. But circumvention also refers to whether someone can attack a biometric system in other ways, such as replaying known good credentials through a network connection.

Benefits of Biometric Systems

The antagonists of data security generally consider any form of security as added complication at best and a violation of privacy at worst. We have instead taken the path that security should be a *business enabler* by adding value in some measurable way. Biometrics are no exception: An organization that is implementing biometrics is doing so in order to fulfill a business objective that is usually tied to the reduction of risk.

The three chief benefits that biometrics bring to an organization are

✔ **More reliable identification:** With biometrics in place, it's far more likely that the person logging in or entering a building is who he says he is. The risk of a lost key card, for example, is greatly reduced when a biometric is required in addition to the key card. And it's highly unlikely that you'll find a finger or eyeball in the parking lot that an intruder can use to enter a building.

✔ **Elimination of *password sharing*:** Because biometrics are associated with a person and cannot be separated from the person, it eliminates password sharing and that satisfies a regulatory requirement for some, and provides greater accountability for all organizations that use biometric authentication.

✔ **More convenient identification:** Depending upon the manner in which a biometric solution is integrated into an authentication/authorization system, the biometric solution may make identification even more convenient than before. For example, Peter can log in to his laptop computer with a swipe of his finger, which takes less time than entering a user ID and password.

Selecting a Biometric System

Different biometrics use different measuring techniques and conditions. Depending on the requirements of a specific access control situation, some biometrics will be more suited than others. For instance, DNA verification would not work well for logging in to a computer system (even one that contains *really* sensitive information), as the DNA confirmation is labor intensive and takes a few days at the very least — by that time you'll forget why you wanted to log into the system in the first place.

The point we're trying to get across here is that any initiative that considers using biometrics is to improve access control requires a *lot* of discussion — as well as the development of formal, written requirements. Failure to take those steps may result in choosing the wrong kind of solution, which can be a very costly mistake.

As with any technology-related project, the very first order of business should be the development of *formal business objectives* that are blessed or (even better) *expressed* by the executives in the organization. An example of a bottom-line objective is to *improve the company's ability to prohibit unauthorized persons from accessing valuable assets*. It's hard to find organizations that set out to acquire the latest technology just because it's cooler than the old stuff in the previous generation. Similarly, nobody adopts the newest technology just to keep the system administrators from getting bored and quitting. (Nobody reasonable, anyway.)

The steps for selecting a biometrics solution are pretty straightforward:

1. **Identify selection criteria.**

   This process includes understanding the physical and logical environments, establishing physical requirements (size, weight, and power requirements of biometric devices, for instance), determining acceptable accuracy and rates of failure (false acceptance and false rejection), and getting an accurate handle on regulatory requirements, budget, implementation effort required, and how soon you need a solution in place.

2. **Identify the field of possible solutions.**

   When selection criteria are established, it will be easier to objectively eliminate unsuitable types of solutions and get closer to a "short list" of candidates. Closer analysis of requirements, features, and budget should help you get to one or two types of biometric approaches that may work for you.

3. **Test potential solutions.**

   When you've eliminated most of the playing field and are down to a short list of specific solutions, you should consider getting hold of an actual product you can test. Most vendors will loan you a reader or two for a couple of months if they think they have a shot at selling you a *lot* of them. When you test a couple of solutions in something close to your real setting, you can see how well they really perform. Be sure you get some users involved in the testing — their observations and feedback will be more valuable than you may realize.

4. **Choose the solution.**

   After you've tested some solutions, you should be able to make a selection. The runners-up will want to know why they weren't chosen; it's best to not burn your bridges, but instead, tell them why (amiably and honestly). Who knows — you might be doing business with them in the future when your needs change.

We describe the entire process of product selection in juicy detail in Chapter 8.

Implementing Biometrics

Once you have selected a biometric system, you'll need to develop a plan to get the system installed, configured, and running. But if it were just that simple, we would not have devoted an entire chapter to the subject. The reality is that implementing biometrics is fairly complicated, not because of the technology but because of the behavioral changes that are required of the people who will be using it and the typically large impact on the organization.

Educating users must begin early and should be structured in a manner that gives them a way to ask questions and express any concerns they may have. We can guarantee that if you're implementing a fingerprint-based system, some of your users are going to express concerns regarding civil liberties and make remarks such as "I'll be damned if I'm going to hand over my fingerprints to my employer!" And we're sure you will also hear something like, "I saw someone pick his nose before using the fingerprint reader — do you think I'm stupid enough to use it too?" Our point is that user education is probably more important for biometrics than for nearly any other kind of IT project. Even switching users from Windows to Linux would be easier, in our opinion — chances are nobody will call you a fascist for doing *that*.

Aside from user education, careful planning is the most important part of successfully implementing a biometric system. If you're going to be installing biometric readers in lots of places, then you'll have the normal logistical challenges of getting equipment installed and configured properly — and making sure the red and green wires didn't get crossed (which could be a problem if your installer is color-blind).

Obviously, we've touched on only a few of the issues that crop up in biometric implementation. For the complete story, turn to Chapter 9.

Understanding Biometrics Issues

Biometric technology is nowhere near universally accepted by all users. There are a number of social and legal considerations that give every organization some pause before taking the jump headlong into implementing a biometric system. Done right, identifying and managing these implications helps an organization make a better decision — not only on the type of biometric technology used, but also on how it will be used.

Privacy

In the United States, Europe, and other regions and nations, citizens have a legal right to privacy — which at times may give the use of biometrics the appearance (if not the fact) of intrusiveness. Sometimes privacy concerns are based on misperception; at other times, they're well founded.

The privacy concern arises primarily because people believe certain biometric data that has been collected by a private organization can later be used in ways that would violate their legal rights — or even cause them more tangible harm. The classic example is fingerprint-based biometrics. When users register their fingerprints, they usually believe that the organization is collecting actual fingerprint images — but generally that's not the case. Typically a biometric system based on fingerprints *scans* the image but stores only a cryptographic hash of the data that describes the print. Hashing cannot be reversed to produce the original fingerprint. Users should feel a little better once they understand this.

Privacy laws

A number of privacy laws in the U.S. provide some vague guidance on the permitted collection and use of biometric data. We say "vague" guidance because most of these laws were written prior to the popular use of biometric technology. Some of the noteworthy laws include the following:

✔ **Privacy Act of 1974**, later amended by the **Computer Matching Privacy Act of 1988**. These laws define how information related to individuals can be collected and used. The long and the short of it is, federal government agencies can collect biometric information and even share it and combine it with information collected from other federal government agencies. However, every citizen has the right to know what information is stored, and must be given access to a process for correcting information that is inaccurate. These two laws don't say anything about how long federal agencies may keep this information, so it's likely they'll have it long after we die.

✔ **Executive Order 12333.** Signed by President Reagan in 1981, this order seeks to encourage the enhancement of biometric collection methods while still retaining privacy. Primarily the order provides guidelines regarding what situations and conditions justify the collection of biometric information, and regarding what types may be collected.

We discuss these privacy laws in more detail in Chapter 3.

Protecting Biometric Data and Infrastructure

Few will argue that biometric data, even if it's obfuscated, hashed, or encrypted, must be protected from unauthorized access, corruption, and loss. Security professionals call this the protection of CIA — the *c*onfidentiality, *i*ntegrity, and *a*vailability of data.

My fingerprints have been breached — can I get new fingers?

Unless you've been living under a rock, you've heard about the scourge of security breaches concerning credit cards, bank-account numbers, and so on. In the case of credit cards, when they're compromised, the issuing bank quickly cancels the stolen card number and issues a new number for the customer.

But what happens if a person's actual fingerprint images are compromised and published? You can't get your fingerprints replaced; they're permanent. Same goes for your iris image and your other physiological characteristics. This fact has driven the designers of most biometric systems to store biometric information in a form that can't be used to derive the original data — not even whether it's a fingerprint, an image of your iris, or a description of the way you speak.

This concern underscores the need for an organization to do an effective job of educating its personnel about the facts — making sure they know exactly how their personal information will be stored and used. Lacking this information, staff members will fear the worst, which can undermine an otherwise-well-planned biometric implementation.

Biometrics are used to protect valuable assets, whether those assets are workspaces containing expensive machinery or computer systems containing sensitive information. The measures taken to protect biometric data should be similar to those used to protect passwords and other credential data.

To understand how to protect biometric data, it is necessary to understand the types of *threats* that jeopardize it. Mostly, these are the same ones that threaten other information assets. There are quite a number of threats that we categorize as natural (such as floods, lightning, and hurricanes) and man-made (such as sabotage, communications failures, and riots).

It is also important to understand the kinds of *vulnerabilities* that may exist in biometric systems and what can be done to minimize these vulnerabilities. Some of these vulnerabilities are the same ones that other types of information systems and networks face — such as exposed cabling, missing security patches, and improper configuration.

The types of *attacks* that can take place against biometric systems include systems attacks, network attacks, application attacks, social engineering, replay attacks, faked credentials, bypass attacks, and enrollment fraud. The first four are the types of attacks that can be launched against any computing environment. The last four are types of attack specific to biometrics themselves — here's a closer look at these:

- ✓ **Replay attacks:** Here, an attacker has found a way to re-transmit known good biometric authentication data over the network in a way that can fool the system into admitting him.

- ✓ **Faked credentials:** This attack uses a forged credential in an attempt to gain access to a system or building. Examples include "gummy fingers" and images of faces or irises.

- ✓ **Bypass attacks:** An attacker may try different methods of breaking in to a system or facility by bypassing the biometric system altogether.

- ✓ **Enrollment fraud:** An intruder may attempt to enroll him or herself in place of a real individual.

For the most part, the tools and techniques used to protect typical computing environments also apply to biometric systems. For instance, all the servers, databases, and network devices that support a biometric system need to be "hardened," have current security patches installed, and have good access management controls in place.

As you might guess, we're hitting the security aspect of biometrics pretty lightly here. The heavy-duty stuff comes later: Chapter 10 has a more complete discussion of keeping biometric systems safe from harm.

Chapter 2

Protecting Assets with Biometrics

• •

• •

*T*o understand how to use biometrics for protecting assets, you must consider what you can actually do with biometrics. Probably the most common use of biometric tools is authenticating known users for the purpose of granting access to sensitive systems or areas; some forms of biometrics are also good at identifying persons whose biometric data is on file without any initial clues as to who they are.

In this chapter, we help you understand when biometrics make sense and start pairing up some kinds of problems with specific kinds of biometric solutions. For a more detailed look at choosing and implementing biometric solutions, take a look at Part III.

Speaking the Language: Biometric Concepts and Terms

For the rest of this chapter to make much sense, it is important that you understand the differences between these concepts.

Authentication implies that someone has presented some sort of credentials or made an assertion regarding his or her identity; here biometric tools can verify that such persons are who they claim to be. The biometric system has a simple task to perform: Collect the biometric information from the user and then compare it to the known biometric information for the person to be authenticated. If the system comes up with a match (within acceptable margins for error), the authentication succeeds.

Identification using biometrics is a somewhat different task: Biometric data is acquired from the subject and then compared to *all* samples in the database, in order to determine who the person is. In the case of a fairly complex but somewhat inexact biometric such as a person's gait — where the margins for error are relatively wide (compared to, say, iris scanning) — an identification system might yield more than one possible match.

Whenever you use biometrics for identification, only the biometric itself is acquired from the subject; the identification is performed without the bias that goes along with comparison to a specific person.

Why not just always use biometrics for identification rather than for authentication? Well, for one thing, it's time-intensive. Identification has to compare each user *with the entire database* until a match is found — far less efficient and less accurate. The "less accurate" part requires some explanation.

Consider what happens if you're using a biometric database of fingerprints — with millions of users enrolled — in identification mode to figure out which user is trying to log in to your computer system. Since the system has to just start at the beginning of the database and start comparing, what happens if there's a fairly close match to the current sample — and it shows up earlier in the database than the correct match? Potentially, the user could end up identified as the wrong person. If you use the same database in an authentication capacity, the user asserts who he or she is, and then the system makes a comparison and decides to accept or reject the assertion. It's still possible to have a false acceptance — after all, the close match is still in the database — but the user would have to choose the *correct* false match from the millions of choices at random. That's unlikely unless the user has direct access to the encoded prints.

All forms of biometrics have error rates associated with failures to identify or authenticate users. Incorrectly accepting a biometric is called a *false acceptance;* the percentage where this will occur for the system or biometric type is called the False Acceptance Rate (FAR). Incorrectly rejecting a biometric is called *false rejection;* the percentage where this will occur for the system or biometric type is called the False Rejection Rate (FRR).

Starting at Square One

We like to start security projects by understanding exactly what business-enabling function the new project is intended to facilitate, and making sure that the project will accomplish what's expected. The number one problem with security projects is that while they're furiously involved in making something secure, they forget that there's an original business need that started the whole process. If you find it difficult to express your security project in terms of enabling organizational goals, it may be time to rethink your organization's actual need for the project.

Recognizing organizational benefit

When we ask security professionals what the organizational benefit of some security project is, we often hear things like "*It will better secure our customer data*" — which is really (believe it or not) beside the point. If the goal of the organization is to secure customer data, we guess we could see that as an organizational benefit, but it's far more likely that the organization exists to provide some *service* to those customers other than just collecting their data and keeping it really secure.

We're not saying that security is unimportant; far from it. The information security of an organization can mean the difference between happy customers who use your services and recommend them to their friends and multi-million-dollar class-action lawsuits against the company for losing all of their credit card numbers. What we are saying is that security isn't a useful goal at the organizational level. Instead, security is a tool we use to manage organizational risks to protect the bottom line.

As Alex Trebek might say if he were moonlighting as a savvy security consultant, "Please phrase your security project proposal in the form of a goal in keeping with the mission objectives of the organization." Okay, maybe he wouldn't say *that*, exactly, but you get the idea. You have two basic ways to look at your deployment of biometrics:

A project whose main point is to use biometrics for something	Typical description: "A biometrics project that requires fingerprint identification for anyone logging in to the customer database."
A project with larger business goals consistently in mind	Typical description: "A project to reduce the cost of data-theft mitigation by introducing biometric authentication for authorized users of the database."

Why is it important to make this distinction? The short answer is that by placing the project in its proper context, you can show others in the organization the value of your project *in terms that make sense to them*. The longer answer is more about how security professionals think and talk about security — and how that affects planning for the organization.

Many security people should consider "catching more flies with honey than with vinegar" when they talk about the job of securing information or things — in this case, the honey takes the form of positive language focused on reaching organizational goals. If your language is inherently focused on the negative aspects of the things you're proposing to "lock out, keep from, exclude," or otherwise stave off, the audience you're most trying to reach may tune you out. That happens even though what you're trying to accomplish is for the good of the organization. So focus your language on what your hearers want to accomplish, using a positive perspective of "providing, enabling, allowing, and making available."

"Wait", you say — "when I'm talking about locking out and excluding, I'm talking about the bad people, not our customers!" Keep in mind that *providing*, *allowing,* and *making available* is the *output* of the organization; the protective stuff is a side effect of good management. Statements that talk about "doing a great job of providing access by making it more convenient and secure" will make sense to management; rants about "keeping the wrong people from getting access by introducing new security measures" will mostly get tuned out.

Understanding your customer

In an organizational context, your "customers" are the internal people who will be using the biometric system to go about a typical day's work. You can apply biometric protections that protect assets to both electronic-access systems (as the equivalent of a password) and to physical assets (in the form of access control). In effect, your internal "customer base" is limited to people who require access to computer systems or secured areas. In all seriousness, the customer base for your biometric security system is twofold:

- ✔ **The organization that requires the kind of security that a biometric system provides.** The organization's goals and requirements should be what actually started the process of securing assets with biometrics, so those shouldn't be a mystery.

- ✔ **The people who actually have to *use* the system you install, probably without much choice in the matter.** In any group of people larger than about three, it's likely that not everyone will be thrilled with whatever biometric solution you choose. For larger numbers, it's likely that the solution may not even work properly for some small percentage of users due to physical limitations.

The key is to understand who the end users of the system will be — both to accommodate the largest possible population and to avoid any gaffes in your technology choices. For example, the building Mike works in also houses services for the blind and partially sighted. The service organization employs several blind people who require access to the building. Although (in most cases) eye-based biometrics would technically work fine, the need to have the user focus on a specific spot to get a good image would rule out retina or iris biometrics for access to this particular building.

Understanding customer needs doesn't just mean understanding the physical characteristics of the population either. While working on a biometric identification system for a large health-care provider (for example), Mike was well down the path of recommending a good fingerprint-identification system when someone told him that lots of people in this environment wear gloves a lot of the time. Oops. Fingerprint biometrics could well be challenging in that situation. Oddly enough, a little testing showed that the print scanners he was interested in could work through rubber gloves just fine — but if they hadn't, and it didn't become apparent until after the system was deployed, it could have been a serious mistake — brought about by ignorance of the daily routines of the doctors and nurses in an emergency room.

Understanding the objective

The objective of the exercise is to *secure the asset*, right? That's super-easy as long as we all agree on what the specific definitions of *secure* and *asset* are. That may sound trivial, but take it from someone who's sat in meetings for hours listening to people debate this very topic — it's more complicated than it sounds, and the complexity crops up the moment you try to apply those definitions to a specific context.

Secure is an elusive quality — unobtainable in its pure form, if you believe most security folks. Humankind has a long history of building walls, forts, and various other obstacles, trying to keep ourselves and our stuff secure — most of which ultimately failed to do so.

Over time, the degree to which you can keep things secure is inversely proportional to how interested others are in breaking that security. For example, the Content Scramble System (CSS) used for protecting DVD movies from piracy lasted for about three years before someone released code that uniformly cracked DVD encryption. Blu-ray was cracked even faster.

Since *secure* seems to be a relative term, it's important — when contemplating a biometric system to keep something secure — to understand what level of security you are really striving for. If you're securing weapons-grade plutonium, we suggest further study (a nice advanced degree program, preferably, and 20 years or so of industry experience), but that's an entirely different scale of project from (say) using a fingerprint ID system to provide access to the company accounting system.

Anti-biometric hobby

For a period of about three weeks in high school, Mike had no fingerprints. Any of his friends and family reading this might be thinking, "I remember no eyebrows a few times from his *explosives* phase, but how did he blow his fingerprints off?" The truth is he actually *sanded* them off through hours and hours of wet sanding, while applying a new paint job to his 1976 Toyota Corona Mark II MX. All those hours of sanding the car also sanded Mike's fingertips to the point where they were tender and had no discernable prints. Fortunately for his authentication needs, prints grow back quickly. But it's worth asking: What processes happen in your organization that might interfere with biometrics, even as side effects?

Imagine that the asset being protected is really valuable — say, a couple of billion dollars' worth of pre-patent intellectual property at a law firm. One of the measurements you might want to consider would be how much someone might be willing to spend to get access to those assets. An attacker might spend a few hundred dollars splicing into the line between the palm scanner and the computer system it's connected to; *voilá* — now somebody can intercept the biometric data and/or replace it with other data. Depending on how well you've protected that piece of wire, a few hundred dollars might be a good investment for access to billions of dollars worth of assets. Specifying a system that encrypts data between the scanning device and the system would mostly eliminate this problem, but it does add additional expense.

Higher levels of security typically cost more money, are less convenient for users, or both. Figure 2-1 graphically illustrates the relationship between convenience and security. If we were to add a third dimension to the diagram, it would be a "cost" axis — because sometimes inconvenience can be overcome by spending more money.

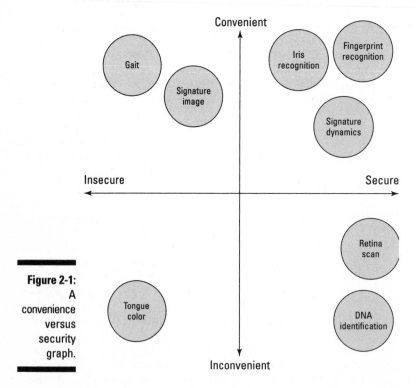

Figure 2-1:
A
convenience
versus
security
graph.

Applying Triple A: Authentication, Authorization, and Accounting

Once you have a complete understanding of the objectives, the customer, and how it all fits into the organization, you're ready to roll up your sleeves and get involved with the real work of biometrics as used to protect assets: authentication, authorization, and accounting (AAA, also known as *Triple A*). If you've heard of AAA before with respect to security, you may have thought of it as the part of a security system that grants or denies access. That's a good start on protecting things using biometrics. Here, we take a closer look at the other two "A"s (authorization and accounting); the idea is to make sure everyone is talking about the same thing when referring to authentication.

Refining ideas about authentication

The act of authenticating someone presupposes that someone's already made an assertion — along the lines of "This is Brenda" — which we are authenticating. Starting out with no information at all is a different task entirely — called *identification*. Biometric systems can be good tools for identification, but if we're talking about securing systems, we want to use the most accurate possible comparisons. That means comparing a new biometric sample to a single existing biometric profile — which is the process of *authenticating* a person. Figure 2-2 illustrates the difference between identification and authentication. It's possible to build identification-based biometric systems for protecting assets, but if you go that route, be prepared to spend additional money to get the level of security that's already available through authentication-based systems.

The assertion itself can be accomplished in many ways, but typical methods include typing in a username, presenting an electronic ID badge, or — in the special case of voice recognition — saying your name. A speech recognition system can translate what you say (your name) into text, and then a biometric voice-recognition system uses the already-collected voice pattern to make the biometric comparison. Note the distinction here between two separate systems: speech recognition and voice recognition. *Speech recognition* translates speech into text, but has no idea who is talking, while *voice-recognition systems* match patterns in the voice to known biometrics.

Because authentication requires an assertion of identity to start, biometric authentication is frequently a *two-factor* (also known as *multifactor*) authentication — the use of more than one item to authenticate the user. In practice, each of the (usually two) factors should be hard to duplicate or obtain; typing in a username isn't a useful factor unless the username is obscure and maintained as a secret. Often multifactor authentication is described as using two, or even three, factors: something you *know*, something you *have,* and/or something you *are*. In a biometric authentication system, we use something you *are* as one of the factors.

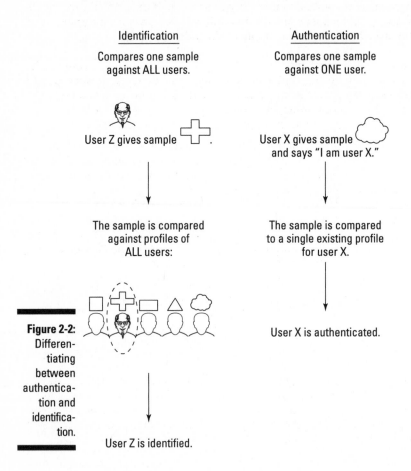

Identification

Compares one sample against ALL users.

User Z gives sample.

The sample is compared against profiles of ALL users:

User Z is identified.

Authentication

Compares one sample against ONE user.

User X gives sample and says "I am user X."

The sample is compared to a single existing profile for user X.

User X is authenticated.

Figure 2-2: Differentiating between authentication and identification.

Authentication provides a mechanism by which a system can verify that users are who they purport to be. The typical uses are for access control to a computer system or other mechanism, or to a physical area. There are other uses for authentication such as verifying that an action was performed by a specific person (discharging a medical patient) or an online banking transaction.

Authorizing actions

Authorization plays a role related to — but distinct from — authentication in securing systems with biometrics. Most security systems make distinctions between the things various people are allowed to do. Authentication allows the system to know who is trying to do something, but *authorization* is how we can conceive of each authenticated user only being allowed to perform the tasks in the system that the rules allow.

Most of the time, the actual biometric system doesn't have any direct role in authorization. That's because authorization is based on rules internal to the protected system. There are a couple of exceptions to that:

✔ At a very basic level, most biometric systems are required to authenticate a user and then grant access to that user based on the successful authentication. The "granting access" part of that process can be performed by other systems, based on information that the biometric system supplies (saying, in effect, *Yup, that really is Rhonda*).

✔ In some cases the biometric system itself may directly grant access based on a successful authentication. Biometric door locks are frequently in the latter category; they can choose to open the door without consulting any other system.

In cases where biometric systems are expected to directly grant access to a physical area or system, the system must understand authorization or everyone in the system will be granted access to everything. In the case of a door lock, for example, you can imagine that on a large campus, not everyone should have access to all doors. Just being in the system and able to get in and out of the lobby and your office doesn't give you access to the shrunken-head collection or the plutonium repository.

Another place where biometric authentication and authorization overlap somewhat is a special case: specifically requiring biometric signatures to authorize actions. In some financial systems, ordinary transactions can happen just because a specific user is logged in to a terminal and going about his or her business; when a transaction over some preset limit is attempted, however, the user is forced to specifically authorize that transaction by providing further proof of identity. Biometrics is a perfect way to accomplish this kind of authorization because biometric data is hard for someone else to imitate.

Billions and billions

While working on a supplemental authorization system for a large financial institution, Mike once saw on-screen what the limit was for triggering a forced re-authentication: The user would be prompted to re-authenticate when any single transaction exceeded nine hundred million dollars. Because Mike knew how frequently the system was being used, he also knew that each of the users in the office where he was working was performing transactions of that size — or greater — at least once each day. That marked one of many times he considered changing his consulting rates on the spot.

Auditing

Audit has a lot of common meanings, none of which are all that pleasant sounding (unless, of course, you are an auditor). In the AAA sense, *auditing* is about having a complete record of the authentication and authorization events. The record of what's been happening is an important part of any system for a lot of reasons — and audit is among them.

In some situations, such as medical applications, a complete record of who has accessed records is more than just a good way of tracking access and behavior; it's a legally mandated requirement, with serious penalties for getting it wrong. Even without legal requirements, having a record of who accessed what asset and when is critical to understanding how people are using your protected systems.

In unusual situations — for instance, when there's evidence of misuse — the audit records are essential to identifying what records may have been accessed and by whom. It's worth mentioning here that audit records associated with data access warrant extra protection. When an experienced system cracker breaks into a computer or access system, the first thing the intruder wants to do is erase any tracks — and the audit records are the target.

You can tamper-proof your audit records by employing these three measures:

✔ Keep your audit records in binary format.

✔ Give the audit records cryptographic signatures.

✔ Archive the signatures separately from the audit records.

These three measures can provide enough protection that an attacker will look for someone else to mess with who's not as careful.

Choosing the Best Fit

Biometric protection systems are uniquely positioned to be the most aggravating, invasive, in-the-way system that your users have ever imagined. After all, the more accurate biometrics require pretty direct personal contact with the system. So a biometric system that isn't suited to the task (or works poorly) is about as bad as information technology gets.

A biometric system well suited to your needs, on the other hand, provides better measures than really long complicated passwords, and saves all that extra typing. Careful thought about how the users work, what their work conditions are, and how they expect to interact with the protected systems or areas will lead to good choices and happy users.

In this chapter, we discuss several aspects of biometric systems that all have some effect on the purchase, use, maintenance, or usability of any such system. It can be tempting to focus on just one or two of these parameters as the most important and pretty much ignore the rest — but resist that temptation. The best approach is to list all the parameters that apply to your situation, assign weights to each of them (based on the specific needs and requirements of your asset protection profile), and see how the numbers come out. Although your installation may have additional items to consider, as an example, we consider security, environment, cost, and regulatory requirements.

Security as a goal

When the goal of a project is to *secure* assets, you might expect that security would always be the highest priority. The reality, however, is that the security of a system must be strong enough to repel only those people who are actually interested in breaking in. Because the required level of security drives costs higher — while simultaneously producing a negative effect on the environment in terms of usability — the security requirements for a biometric system should be carefully considered with an eye toward containing cost *and* providing a workable environment.

For example, even if we are just limiting our scope to fingerprint biometrics, there are security considerations to take into account when choosing among the systems available:

- **Secure transmission:** Is it possible to intercept the data coming from the reader? Is the wire protected? Is the information encrypted?

- **Secure storage:** How safely is the biometric database stored? Is the information in the database subject to abuse if stolen, or is it useless outside the system?

- **Secure design:** Does the biometric sensor itself allow latent prints to be collected (glass surface scanners).

Depending on the security required for your installation, you may need to consider points like these.

Environment as a factor

When you're fitting a biometric system to your setting, you have to take into account not only the working environment as it exists prior to installation, but also the *expected* working environment after adoption of the system. In a perfect world, the biometric system actually improves the working environment by providing a more convenient authentication and authorization method, saving users time and making their jobs easier. Since we rarely live in a perfect world, sometimes the main reason for instituting a new biometric system is to usher in a higher level of security — and some level of inconvenience is expected as a part of that process.

The introduction of a new way of doing things is almost always problematic in the short run; everyone has to get past the initial retraining before they can begin to use the new system efficiently. Sometimes the new system has to satisfy additional security concerns or regulatory-compliance requirements — and so will continue to be a pain to use, even after the initial break-in. The goal, in those cases, is to meet the new challenges while minimizing the new system's negative impact on the working environment.

One great way to understand each system's possible impact on the working environment is to observe working installations that use the technologies you're interested in — either in place somewhere else or as a test environment you build yourself. (For a more detailed discussion of building a test environment and getting good data from it, take a look at Chapter 8.)

Cost as a limitation

Of all the parameters you may need to consider and balance in choosing the right biometric technology for your project, *cost* is possibly the only one you'll always try to push in the same direction — downward. While money can't buy everything, it nearly always influences the other parameters in a specific direction. For example: Higher levels of security can be purchased, as well as sensitive, high-resolution biometric sensors that reduce the FRR or allow acquisition of biometric information from a distance (reducing the impact to the working environment). But it'll cost you.

Because cost is nearly always an issue, the challenge is to buy as much security, integration, and regulatory compliance as you really need — without breaking the bank. You need to figure out how much security is enough. About the only thing you have working in your favor with respect to cost is time. As with almost any technology, the longer you can put off your project, the more you will be able to buy (because the cost will go down over time) — which means you'll accomplish more of your objectives with the same amount of money. Technology purchases can be one of those times when procrastination pays off.

Regulatory considerations

We feel that regulatory compliance is a useful example because accomplishing compliance goals is really an absolute, black-or-white goal. Cost, environment, and security are sliding scales; compliance-or-noncompliance is pretty clear-cut. If your organization must comply with regulations that describe how authentication and authorization are used, then you really don't have much wiggle room in this area. Your new project will meet your compliance goals, or it will not be useful to the organization.

The absolute nature of this parameter can drive other parameters (such as cost) without providing any opportunity for compromise. If you must achieve a specific FAR, and the only available biometric system that can do the job costs twice your expected budget, you'll be spending the extra money or risking fines from the regulatory body. So you must ask which will cost more — and make your choice based on the answer.

Regulatory compliance is specific to the regulatory body (typically federal, state, or industry based) and is often more art than science. The problem with compliance to regulations at the federal level is that the federal government is not in the business of advocating specific technologies and wants to make sure that regulations are not so specific that they need to be changed often as technology advances. Due to these limitations, regulations tend to be a bit vague about technology or specific methods and rely on industry best practices to guide people.

For example, a regulation might say that your authorization system must prevent *repudiation* of authorized actions. In other words, a user must *not* be able to say she did not authorize an action that she really did authorize. Most biometric systems offer some level of non-repudiation because they tie the authorization to a physical or behavioral aspect of the user, but iris biometrics offer quite a lot stronger non-repudiation than simple signatures.

Chapter 3

Biometrics in Our Lives

. . . .

In This Chapter

▶ Exploring privacy issues

▶ Looking into the legal side of biometrics

▶ Investigating ethical issues

. . . .

Generally technologists haven't done such a great job anticipating the impact to society of any particular new technology or platform. Partly, that is just the nature of the game. If you give a child a plastic dinosaur in a box, it's hard to predict whether the box itself will eventually become the command module for a spaceship trying to divert the giant asteroid and save the Cretaceous denizens from their horrible fate. As a society, we behave in some ways exactly like a curious child when it comes to technology. We explore every possible way we might use the new tool — some of which were never envisioned by the tool's creators — and we keep pushing the limits of every idea until it no longer resembles the original concept. We (your authors) sincerely doubt that the estimable visionaries at DARPA were thinking about the kind of revolution e-commerce would eventually bring to the world in the form of the worldwide Internet.

Biometric technology is another one of those tools that evolves this way. As a base concept, it's simple: Use recorded knowledge that describes some aspect of us (either physically or behaviorally) and then use that knowledge to later identify or authenticate the person who has those attributes — say, you. In order for your biometric measures to be really useful, the information gathered must be detailed and unique to *only* you, and clearly associated with your identity in whatever system is making the identification or authentication. That's where it gets complex.

This chapter is all about the vast implications of such a simple concept when we consider collecting highly personal information from people — of a sort that in many cases will never change over their lifetimes — and then using that information in government, at work, and in the home.

Privacy Issues

Any time we consider giving up personal information to someone else, we have to consider what the implications are. For example, we might be perfectly comfortable with telling family members where the secret stash of emergency money in our home is, but not as willing to share that information with a co-worker. These days there is so much information about each of us that seems to just somehow make its way to the Internet, government offices, and employment databases that savvy information users are being very careful about who gets access to what.

Biometrics are at once both very personal and potentially damaging if somehow used in ways we did not expect when providing the information. Also, since the act of collecting biometrics involves observing behavior or scanning some body part in detail, it can seem even more invasive than it really is. For all these reasons, it makes sense to understand the issues surrounding privacy and biometrics.

Constitutional privacy protections

In the United States the right to privacy is reasonably well established and constitutionally based. Although the U.S. constitution does not explicitly name privacy as an individual right, interpretations of the First, Third, Fourth, Fifth, Ninth, and Fourteenth amendments by the Supreme Court uphold privacy as a constitutional right; specifically, a right to physical, decisional, and informational privacy.

There are potential problems with each form of privacy recognized by the Supreme Court, but in most cases (as we explain) the concern over biometrics derives from issues of informational privacy. To understand why, we look briefly at what each form of privacy means — and give an example of how each might be of concern in biometrics:

- **Physical privacy** concerns itself with freedom from monitoring by or physical contact with others. Although this might initially seem to be a potential clash between biometrics and your privacy, such a conflict is unlikely unless someone forces you to provide a biometric sample, or takes pictures to be used in this way when you're in your own home or another nonpublic place.

  Although there haven't been many challenges of an individual's physical right to privacy with respect to biometric information, we could imagine challenging a police officer taking a high-resolution picture of Mike to capture his iris image for the purposes of biometrically verifying identity on Fourth Amendment grounds.

✔ **Decisional privacy** is the freedom to make choices in personal matters. There are numerous emotionally charged examples of the Supreme Court decisions in this matter, but to stay out of those fights (most of them, anyway), imagine a state passing a law that prevents you from getting a tattoo. The decision to get a tattoo is a personal one, arguably affecting nobody else, and any such law would be found to be unconstitutional.

As with physical privacy, there are few good examples of this with respect to biometrics in real life, but one example might be the government requiring Mike to submit his iris information when he really doesn't want anyone to know what color his eyes are. Requiring Mike to lift his dark glasses and stare into the camera might be considered a violation of his decisional privacy. Such a measure would almost certainly be challenged as a violation of informational privacy (which we look at next) as well.

✔ **Informational privacy** is far easier to understand and to find examples for. Informational privacy is the freedom to control access to information about oneself. Since biometric information is always tied to information about the individual, subsequent use of that information has the potential to violate this privacy right.

For example, if Mike submits his fingerprints to an employer to authenticate him as he arrives for work, and that employer then uses those prints to run a criminal background check, the employer has most certainly used the biometric identifier in a way that might violate Mike's informational privacy rights.

Note our use of the words *might*, *may*, and *possibly* in the preceding paragraphs. We're not trying to be imprecise on purpose. The problem is that law, especially constitutional law, is whatever the courts interpret it to be at the time. Until we see court cases establishing those interpretations, it's impossible to know for sure what the outcome will be. Also keep in mind that privacy rights are not absolute rights — and are frequently weighed against the good of society. Your right to physical privacy does not extend to hiding contraband on your person and smuggling it out of the country, for example.

Statutory privacy protections

The Privacy Act of 1974, as later amended by the Computer Matching and Privacy Act of 1988 (the act), is by far the most interesting statutory description of privacy and rights to privacy in the U.S. A complete description of this act and its implications would be the topic of several books this size, but we can cover most of the ground important to biometrics pretty quickly. The 1974 act, as amended in 1988, can be found on the U.S. Department of Justice Web site:

`www.usdoj.gov/oip/privstat.htm`

The act pertains to the collection, use, and dissemination of records maintained on individuals as a part of a system of records. More simply, the act defines how the federal government can collect and use information about its citizens. On the face of it, that sounds perfect: a single place where all our privacy questions are answered with no ambiguity. Those of you who've actually *read* an act of Congress can stop laughing now, and we'll get back to the serious business of trying to understand this act.

Defining terms

Even though the act starts with 13 definitions that cover everything from "individual" to "record" and "system of records," most of the time spent in court discussing the act is spent arguing about what each definition really means. We're not going to discuss all the definitions here; most of them are pretty boring and won't add much to the discussion. We present the definitions for *record* and *system of records* here, though, because it's critical to understand those terms to understand the current discussions regarding the act and biometrics. What follows is the exact wording from the act, with the numbers removed that denote which definitions these are (numbers four and five) and capitalization changed for the beginning to improve readability.

> The term "record" means any item, collection, or grouping of information about an individual that is maintained by an agency, including, but not limited to, his education, financial transactions, medical history, and criminal or employment history and that contains his name, or the identifying number, symbol, or other identifying particular assigned to the individual, such as a finger or voice print or a photograph;

> The term "system of records" means a group of any records under the control of any agency from which information is retrieved by the name of the individual or by some identifying number, symbol, or other identifying particular assigned to the individual;

The definition of *record* includes the term "finger or voice print or photograph" — so it seems a slam dunk that a biometric detail would automatically be considered as a record for these purposes. While doing background reading for this book, however, we found arguments that question the validity of using such biometric details as records for the purposes of the act. That's because a given detail may not contain an individual's "education, financial transactions, medical history, and criminal or employment history." These analyses seem to leave out the phrase "including but not limited to," which comes right before that list.

It seems clear that any useful biometric detail (which we call a *record* from now on) will also include information that groups it together with other information about the individual. A fingerprint that doesn't have any associated data is unlikely to be of any use. It's possible to create a situation in which the definition or record does not apply, but more than likely any such biometric records will always be a part of a "system of records" that links the biometric details to other information about the individual.

What it all really means

Assuming that the act does apply, in almost all cases, to biometric information collected by the U.S. government, here's the main question: What does the act say about what can and cannot be done with that information?

In essence, the act tells federal agencies that they may not disclose any record that is contained within a system of records to any other person or agency — not without the written permission of the person to whom the record pertains. Then the act goes on for a while making exceptions to this for Congress, the census, the National Archives, law enforcement, and the Comptroller General — and a general exception from the 1988 amendment, allowing agencies to combine information and use computer matching to compare one record to another. Aside from those specific exceptions, they can't share biometric information with anyone.

Other important things to note is that each agency that collects such information is required to allow individuals to see the record pertaining to them, and to request changes to that record if they feel there are inaccuracies. The agencies are also required to keep detailed records when disclosures are made in accordance to the many exceptions listed.

The difference between having and using

One last note about the act is that it does not prevent any agency from collecting just about any kind of information about you that they choose. The act is strictly about what they can do with the data once they have it. Something missing from this act entirely is anything describing how long such information can be kept. Since computer storage seems to keep getting cheaper and faster, it wouldn't be a surprise if our biometric data in the hands of the U.S. government outlives all of us.

Executive Order 12333

In the final category of U.S. government documents that have some bearing on our privacy and how the government handles biometric information, we have Executive Order 12333 signed by President Ronald Regan on December 4, 1981. Executive orders are not laws, and do not have the force of law. What power they do derive is from their typically being in support or clarification of current laws and act as a catalyst for other members of the executive branch to carry out the wishes of the president. Both Congress and the Supreme Court have the ability to effectively nullify or overturn an executive order — which the Supreme Court has done twice in the entire history of the U.S.

This particular order's purpose was to encourage U.S. intelligence agencies to enhance information-collection techniques and methods, while still maintaining a balance with personal privacy and freedoms of American citizens.

This directive permits collection — for intelligence purposes — of information on American citizens as long as the information meets the following criteria (as laid out in the directive):

(a) Information that is publicly available or collected with the consent of the person concerned;

(b) Information constituting foreign intelligence or counterintelligence, including such information concerning corporations or other commercial organizations. Collection within the United States of foreign intelligence not otherwise obtainable shall be undertaken by the FBI or, when significant foreign intelligence is sought, by other authorized agencies of the Intelligence Community, provided that no foreign intelligence collection by such agencies may be undertaken for the purpose of acquiring information concerning the domestic activities of United States persons;

(c) Information obtained in the course of a lawful foreign intelligence, counterintelligence, international narcotics or international terrorism investigation;

(d) Information needed to protect the safety of any persons or organizations, including those who are targets, victims or hostages of international terrorist organizations;

(e) Information needed to protect foreign intelligence or counterintelligence sources or methods from unauthorized disclosure. Collection within the United States shall be undertaken by the FBI except that other agencies of the Intelligence Community may also collect such information concerning present or former employees, present or former intelligence agency contractors or their present or former employees, or applicants for any such employment or contracting;

(f) Information concerning persons who are reasonably believed to be potential sources or contacts for the purpose of determining their suitability or credibility;

(g) Information arising out of a lawful personnel, physical or communications security investigation;

(h) Information acquired by overhead reconnaissance not directed at specific United States persons;

(i) Incidentally obtained information that may indicate involvement in activities that may violate federal, state, local or foreign laws; and

(j) Information necessary for administrative purposes. In addition, agencies within the Intelligence Community may disseminate information, other than information derived from signals intelligence, to each appropriate agency within the Intelligence Community for purposes of allowing the recipient agency to determine whether the information is relevant to its responsibilities and can be retained by it.

The important thing to consider from the order with respect to biometric information is what can be considered "public information." Is anything that can be collected in a public place, such as your face, your gait, or your irises public information? If so, this presidential order makes all of that fair game for intelligence agencies to collect regarding American citizens. To be fair, there is language in the order that requires compliance with the law and the Attorney General — but it seems very unclear (based on this order) about what biometric information intelligence it allows agencies to collect. If you are interested in reading the original order, you can find a copy on the Central Intelligence Agency's Web site:

```
https://www.cia.gov/about-cia/eo12333.html
```

Although some executive orders have built-in expiration dates, 12333 does not — so it's still in full effect. Unless Congress provides active leadership to specify what information can and cannot be collected on U.S. citizens without their permission or knowledge, this executive order effectively defines the position of the executive branch.

Too much information

Many of us would like ways to identify ourselves for access to work, commerce and even our homes, without the bother of having to remember dozens of passwords for each environment. What could be simpler than using a technology that recognizes something unique about us and applies that information for this purpose? After all, when we're performing these various identification and authentication tasks, the only thing we never leave behind or forget is ourselves.

Many of the biometrics techniques we discuss in Part II of this book gather information about us that might be more revealing than intended. A simple example: A grocery store that's using fingerprint scanning to authenticate payment for groceries would know if you were missing (say) the index finger on your right hand.

It's possible that you don't care about the grocery store knowing something like that, but it's certainly very personal information that the store generally has no reason to possess, but cannot help knowing in this scenario. Even if you're comfortable with your corner grocery knowing this, will you still be okay with it when the corner grocery is bought by (say) the giant conglomerate MegaFudCorp International?

The point is that it's impossible to gather information that is by its nature personal to us for biometric purposes without also potentially gathering irrelevant *but possibly sensitive* information at the same time. Some biometric technologies — such as iris and retinal scanning — even double as medical diagnostic tests, since changes to these body parts typically happen only as a result of direct physical damage or certain diseases. Are you okay with medical information in the hands of people not involved in your health care?

Even behavior-based biometrics might reveal the early onset of some neurological disorders since we tend to choose behavioral characteristics that are so ingrained that major differences in (say) gait could be an indicator of a tumor or other neurological event.

The real problem here is that for some events, there's no practical way to mask the irrelevant personal information. If your company uses retinal scanning to allow access to the server room and you start failing to authenticate, even after re-enrolling a few times, there are only a limited number of explanations for that — nearly all of which include a disease that messes with the fine capillary structure of your retina. Even if the system itself isn't recording these events (which would be odd, since they're security related), the humans who would have to assist in reenrolling you would quickly come to the conclusion that you're ill — and we have no simple way of deleting that information from their brains. Result: The company or people in the company know about a medical condition that they have no right to know about. Biometrics are going to challenge lawyers and ethicists — especially with regard to the use and protection of what the Health Insurance Portability and Accountability Act (HIPAA) calls EPHI (electronic protected health information).

Protecting biometric data

To return to the grocery example, do you trust the management of the great big MegaFudCorp to safeguard your personal information? Well, we give such companies our credit- or bank-card information all the time — so what's the big deal with a little bit of biometric data? The difference is that you can *cancel* your credit card, but your unique biometric data changes slowly (or doesn't ever change) — and once someone has it, there's no way to make it invalid. Although it's easy for organizations to acquire information, it's quite another matter to completely purge that same information.

This concept is really at the heart of several battles over biometric identification systems proposed everywhere. From the perspective of the person or company collecting biometric information to identify or authenticate you, it's the same as a password or a challenge response. (You know the typical gambit: "What's your mother's maiden name?") For you, it's potentially a key to your finances, front door, and medical records all rolled into one — which can never be recalled or changed if it's stolen. No surprise if you consider this data critical to protect — but it would also be no surprise if MegaFudCorp didn't consider it nearly that critical.

Proponents of using biometrics have good arguments for why this isn't as big a problem as it may sound — especially for the most widely used technology, fingerprints. Fingerprint-identification systems almost never actually *store* a fingerprint in its entirety; instead, they only store a few data points that correspond with the representation of minutiae that the system chose as best for identification purposes. Since you don't have the actual picture of a fingerprint, the theft of the fingerprint data isn't a problem, right? With fingerprints, a hash of the print data is all that's really required for authentication, but identification really requires the whole print to be available since a one-to-many match can require additional analysis. A *hash function* takes relatively complex information (like your fingerprint information) and turns it into an integer (the hash) which can be used as an index — in this case, an index into a biometric database.

Without the original fingerprint image, you couldn't re-create a fingerprint that would fool a human examiner for long, but theoretically you could create a bogus print that would fool the specific system that collected the data. Since you know what it's looking for, that's all you really need to re-create.

For most of the other forms of biometric information, a lot more detail is captured and stored, but they are in turn far harder to imitate or falsify. For those forms, though, possession of the information is the direct harm to your privacy. There's no good reason anyone that you haven't shared it with should know about the vein structure of your hand or the metal pin in your index finger.

Keeping a check on Big Brother

Easily the largest biometric projects in the world are all government sponsored and operated. Although it's possible that you may completely trust your own government to never misuse personal information, it strains credibility to imagine that you might trust other national governments to do so. We don't care where you live, or what government we are talking about, at some time in history it has likely misused personal information, violating the privacy and civil rights of its own citizens, and has an even worse record regarding the personal privacy and private information of noncitizens. Both authors of this book are American citizens with long histories of patriotism and cooperation with law enforcement, so when we bring up events such as the internment of Japanese Americans during WWII and recent questions about warrantless wiretaps, it's with the words of Andrew Jackson's farewell address in mind,

> *"But you must remember, my fellow-citizens, that eternal vigilance by the people is the price of liberty, and that you must pay the price if you wish to secure the blessing. It behooves you, therefore, to be watchful in your States as well as in the Federal Government."*

We're not knocking any particular government here, just saying that we get worried when anyone creates a database to track the citizenry.

Biometric passports

As of early 2008, at least thirty-nine countries have passports with biometric information that is contained on a chip embedded in the passport and is readable wirelessly. In many cases, the only biometric information currently on the chip is a photograph, which is also represented non-electronically in the passport. In some cases, however, fingerprints are also captured and stored on the passport as well. Proponents of biometric passports say the additional information contained on the card allows ports of entry to know for sure that the person in possession of the passport *is* the person it was issued to — since the information can be compared directly to biometrics gathered while customs officials watch. In theory, the use of cryptographic signatures ensures that the data has not been tampered with and the passport is valid.

As a basic rule of thumb, whenever you hear someone make an absolute claim about security (such as "This will keep your data safe"), always assume that somebody is either ignorant or lying. The best encryption algorithm in the world will *not* keep your data safe if it's not implemented properly. Some of the encryption used for passports has already been cracked due to poor, nonrandom key choices and other flawed implementation. Assurances that you have to be "very, very close" to someone's passport to read its data also went out the window the 2006 Black Hat conference in Las Vegas as people were able to demonstrate reading passports from about 2 feet away. It's also been demonstrated that RFIDs containing malicious data can be used to disrupt the devices that read the data in passports — in essence, a Denial of Service attack on a port of entry.

Oddly enough, if you're carrying around a biometric passport in your pocket, we don't need to use biometrics to track your movements and identify you. All we need to do is grab the encrypted blob off of your card (apparently pretty simple to do) and then compare the blob we just captured as you walked in the door with the one we have. We don't have to decrypt it, we just compare the encrypted data and if it's the same, we know it's the same passport.

We aren't particularly worried personally about the government having a copy of our fingerprints as we are always careful to wear gloves when we're doing something we don't want the government to know about. We do worry about *any* government having a complete database of biometric information because we don't see that level of tracking as a function of government. The temptation to use such a database in new and interesting ways will always be there, and easily justified as protecting society from a very real threat.

Other repositories

As biometric systems become more economical and simpler to use and administer, we'll no doubt see many organizations that need to verify identity offering — or demanding your acceptance and compliance to — new schemes. Banks are already using fingerprint, iris, and retinal data to authenticate transactions; some look at fingerprint readers that fit into a credit card to show that you were holding the card at the time of the transaction.

The United Arab Emirates has deployed an iris-based identification system that compares every person entering the country to a database of several hundred thousand deportees. This system has been wildly successful in achieving its objective of not allowing these deportees back into the country — and has helped catch and arrest more than 73,000 returning deportees. To accomplish this, all the UAE government had to do was capture and compare the irises of every single person to enter the country. Although the current system does not apparently keep a copy of (say) your iris biometrics, it's pretty easy to imagine that as a next step to further protect the population.

Maintaining the balance

We live in a world where sometimes our privacy is in direct opposition to our safety and ability to counter threats to society. If law enforcement knew the exact location of every person at all times, we guarantee that the crime rate would drop — but we don't want that much intrusion into our daily lives and activities. A mandatory archive of all fingerprints and palm prints of every citizen would also be tremendously helpful in apprehending criminals when police find prints at a crime scene. That, too, seems to be too large an invasion of our privacy to be allowed.

Official government databases are by no means the only threat to our privacy; employers, banks, and retailers are beginning to gather biometric information for their own use. Let's say the police have good reason to believe that one of your fellow employees robbed a bank yesterday because the clerk recognized your company logo on the bag that person stuffed the money into, and the logo showed up on the shirt of the masked robber. Since the police don't know exactly which employee it was — but let's say they know that your company identifies employee movement through the factory by gait-based biometrics — they get a warrant for *all* gait measurements so they can use them for comparison to the video taken by the bank cameras. Result: Even though you, of course, were not a part of the crime, the police still have some of your biometric data. The police might later decide to compare that data to crowd footage of the latest anti-asparagus rally, just so they know whom to look out for in future asparagus-based confrontations.

It's clear that a healthy balance between individual privacy and the safety of our society is important, but given the ongoing debates, it's equally clear that we don't all agree on where that balance lies. In the U.S., we tend to err on the side of privacy and individual rights where we can — sometimes even at the expense of the general welfare and the crime rate. We do this because we feel that the loss of our freedoms and ability to go about our lives without official interference is more important than the threat posed by people who use anonymity to strike against us. Figure 3-1 illustrates the concept of a "sweet spot" where privacy, security, the will of the public, and invasive knowledge of personal information are balanced.

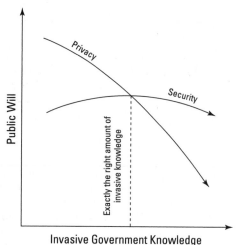

Figure 3-1:
The graph
of mythical
balance.

Biometrics and U.S. Law

In the privacy discussion earlier in this chapter, we cover the basis for the privacy of biometric information in U.S. constitutional and statutory law. In this section, we examine some of the specific new laws and programs that are directly concerned with biometrics.

U.S. and state laws form a large, dark, and sometimes completely inexplicable labyrinth of twisty passages that yield fertile ground for legal scholars, judges, lawyers, and dozens of other professionals who don't always seem to agree about the correct interpretation, scope, or jurisdiction of any particular paragraph. Since we are none of those people, that means we're going to miss lots of laws that we don't know — and get some of it wrong that we think we do know. We promise to feel momentarily contrite for about 30 seconds on the first Sunday after the publishing date of this book to compensate. Synchronize your watches!

We discuss two categories of law with respect to biometrics:

- ✔ Laws that are written specifically to address biometrics issues and needs
- ✔ Laws that have some effect on the collection and use of biometrics

The widespread use of biometrics is relatively new, so few laws directly address their use and misuse. There are lots of laws regarding the collection and use of personal information that were not written to address issues in biometrics — but they certainly apply to the kinds of information we're collecting, and to how we use that information.

Passport and entry

Prior to the tragic events of September 11, 2001, the only biometric information collected and used at U.S. ports of entry were the photographs on passports. These photographs were used primarily to perform a manual check of the passport photo, comparing it to the person presenting the passport to verify that the person entering or leaving the country was indeed the holder of the presented passport.

Although not explicitly biometrics-based, new U.S. passports contain a digitized picture of the holder, which can be used for facial recognition if the need arises. Speaking of need, 64 kilobytes of storage are available on electronic passports — which could be used to store additional biometric information — also if the need arises. Figure 3-2 shows the e-Passport symbol. Although some privacy advocates might disagree, it's hard to imagine a worse scenario for biometric use of a passport than facial recognition alone. If biometric information is to be stored on these documents, it would be nice to use something with a lower rate of false acceptance and false rejection than you get with facial recognition alone.

e-Passport
symbol

Figure 3-2:
e-Passport
symbol.

Those eyes

In 1985, a photograph by Steve McCurry of a young Afghan woman with sea-green eyes appeared on the cover of *National Geographic*. Although you may not have read the accompanying article, the photograph is iconic and well known throughout the world. The penetrating gaze from a young woman who had already seen a hard and dangerous life is disturbing and lovely at the same time. Seventeen years later, that original photograph was used to verify using iris biometrics that the woman was Sharbat Gula, whom McCurry had located again by returning to the same region and doing considerable detective work. The resulting match has a one-in-one-hundred-million chance of being incorrect. We discuss this incident further in Chapter 13.

The events of 9/11 caused the U.S. to reassess how we think about border security — and to pass laws requiring new biometric collection and verification systems at the border. The USA Patriot Act Title X required the Attorney General to study and report on the feasibility of using fingerprints to identify people as they enter the U.S. This study resulted in the US-VISIT (United States Visitor and Immigrant Status Indicator Technology) program.

US-VISIT is a program that starts with the collection of fingerprint biometrics, sometimes at the point of entry and in many cases overseas before flying to the U.S. and then allows custom officials to compare the biometric information to 3.5 million people on watch lists whom the U.S. would like to prevent from entering the country. As of June 2007, the program had collected fingerprints on 100 million people — which will be archived for seventy-five years before being discarded. Initial implementation of US-VISIT collected only the index-finger prints from each hand, but these are quickly being replaced by systems that capture all ten fingerprints.

The transition from two prints to ten, according to official statements by the State Department, will make the US-VISIT program match international standards and provide additional information to reduce mismatches. Although this is a true statement from the State Department, the real value of ten prints versus two is more subtle and important to note: Collecting and comparing two arbitrarily selected prints (each index finger) is used primarily as an authentication of known identity. In this case, however, you're comparing a known collected print to a new sample taken from a person you want to authenticate, and seeing what you get. With all ten prints, you can start thinking more about identification in addition to authentication. Here's an example . . .

Imagine that you're an agent working for Interpol, and you discover a usable fingerprint in the fragments of a bomb casing. You save the print and compare it to known prints but come up with nothing. Being a good agent, you supply this print to countries and agencies that would have an interest in helping you find the bomber, and you hope for the best. Two months later, U.S. Customs stops your bomber at the border because the print turns out to be the right pinky print of someone they just scanned into the system in Los Angeles. A two-print system would have missed this person completely since the print wasn't from a finger (the index finger) that the system cared about. By collecting *all* the prints, the system practically guarantees identification of any individual whose fingerprint has ever been entered into the system by any means — within the capabilities of the fingerprint system itself.

Along with the fingerprints, a photograph is captured electronically that records the face as further information about the individual. Although a full-face photograph is by no means optimal for iris collection, a good-quality photo has been used for positive identification using iris biometrics before.

Special jobs

You'll never get anyone in the U.S. Department of Homeland Security to admit on the record that they would like to have a rock-solid, biometrically enhanced identification card for every U.S. citizen — and capture iris, retina, and fingerprint scans at birth. You can bet a lot of folk there think that way, however, based on the number of citizens required to furnish biometric information for their government or security-related jobs. It's not politically wise to talk openly about tracking U.S. citizens with sophisticated biometric technologies at a national level, so what seems to be currently happening is a creeping requirement of biometrics into specific or special-purpose IDs for small segments of the population (that seem to increase in size daily).

Transportation Worker Identification Credential (TWIC)

Congress, through the Maritime Transportation Security Act, directed the issuance of a biometric credential to

> *"...any individual with unescorted access to secure areas of facilities and vessels and all mariners holding Coast Guard issued credentials or qualification documents."*

By U.S. Coast Guard estimates, that's about 1.5 million workers — including longshoremen, truck drivers, merchant mariners, and port employees — nationwide.

Enrollment for this card includes the collection of a full set of fingerprints and a digital photograph, and it is used to identify and authenticate these workers in areas where a person might be able to effectively assist or perpetrate smuggling or generally violate customs laws.

Homeland Security Presidential Directive 12 (HSPD-12)

HSPD-12 provides for a common identification standard for federal employees and contractors using two fingerprints and a photograph. Since this directive encompasses all the agencies of the U.S. government and approximately 1.8 million employees and contractors, actual enrollment and issuance of these cards has been glacially slow. The implementation and administration of the rollout has been left in the hands of the individual agencies, each of which has its own ideas about the best way to accomplish the task.

Unregulated multitudes

While doing background research for this section of the book, it became clear to us that many professions require practitioners to be fingerprinted — but are not necessarily using those fingerprints directly as a biometric measure. In such cases, biometric authentication and identification is just about installing the technology; the data-collection part has already been accomplished. Since no central authority is keeping track of these unregulated fingerprint collections, there is no good measure of how many people fall into this category, but it's pretty easy to estimate it in the tens (or even hundreds) of millions of people, since some of the categories are quite broad. A partial list of these include lawyers (requirement for admission to the bar in most or all states), New York City welfare recipients, criminals, private investigators, and many law enforcement officers.

Everyone else

Although it's still a political hot-potato, there are a couple of initiatives that could enroll every U.S. citizen in a biometric identification system.

REAL ID

The REAL ID Act of 2005 (which is division B of the Emergency Supplemental Appropriations Act for Defense, the Global War on Terror, and Tsunami Relief, 2005), requires people who want to open bank accounts, board airplanes, and enter federal buildings to present identification that meets certain standards. Not too surprisingly, REAL ID has provisions for including biometric information, but currently does not require it as a part of REAL ID-based driver's licenses (unless you count the photograph, which can be used for face recognition). As things stand now, people born after December 1, 1964, will have a REAL ID by December 1, 2014, and all others by December 1, 2017.

Not surprisingly, many states have raised numerous objections to a system that dictates rules in an area where states have had carte blanche: driver licensing. The rules control what can be used for proof of identity, security of the physical offices where REAL IDs are issued, and background checks for the employees of any department that can issue licenses. These additional rules have the potential for adding lots of additional wait time to the process of obtaining a driver's license, which already has a reputation for being an interminable process.

Crime and punishment

One category of biometric data collection in the U.S. that we don't cover much here is actually one of the largest (and certainly the oldest) collection — that of criminals. If you're convicted of some crimes, or even suspected in some cases, your fingerprints are collected (with or without your cooperation) and kept on file. These files are shared pretty widely among law enforcement organizations and even around the world in some cases.

Oddly, the personal information of people convicted of crimes is not very well protected in general. In one case, a file with information about persons on the FBI's Most Wanted list included the usual names, aliases, and other personal information — as well as *their correct Social Security numbers*. The consequences of this inclusion would be somewhat amusing if it weren't for the potential waste of law enforcement resources. An unsuspecting identity thief is in for the surprise of his life when he starts impersonating an FBI Most Wanted fugitive.

Protections against misuse

While doing the background research for this part of the book, it became clear that an exhaustive list of the places U.S. citizens were required to submit biometric information for identification is a much larger task than you might think it would be. If we include DMV (Department of Motor Vehicles — the State agencies that issue driver's licenses) photos as facial-recognition biometrics, the number jumps to equal nearly the entire adult population — and since DMV photos are used for this purpose, that's not a big leap. If we limit the list to more accurate and unique means such as fingerprints, the number drops from hundreds of millions to tens of millions — for now.

With all that biometric data lying around in government offices, workplaces, and computers belonging to random merchants, you'd think there would be lots of laws about how this information should be protected and how long anyone is allowed to keep it. As is typical for emerging technology, however, the law hasn't really caught up to current practice yet. The U.S. has no laws regarding how long someone can keep biometric information once it's collected. Once you hand someone your biometric keys-to-the-kingdom, that person or institution may keep those keys indefinitely.

Protecting your biometric information

Lack of protection for personal information — and the resulting personal, privacy, financial, and legal mess — is in the news almost weekly these days. All it takes is for some entity with possession of your Personally Identifiable Information (PII) to leaves it accidentally on the seat of an airplane, in the trunk of a rental car, or sitting on an unpatched Web server where someone strolls/surfs by and grabs it with very little effort. As of early 2008, there are (conservatively) 200 million records containing sensitive personal information exposed in this way in the U.S. alone.

With all the dangerous activity and the plethora of entities that might collect sensitive information, our legal system finally reacted. On July 31, 2003, the state of California enacted a security-breach notification law — the first of its kind. As of February 2008, 38 other states have adopted similar laws, most of which are based on the groundbreaking California law. Table 3-1 shows the actual adoption rates for the individual state laws, and how even something so obviously beneficial takes many years to reach widespread adoption.

Table 3-1	Data Breach Disclosure Laws
Effective Date	*State*
July 1, 2003	California
March 31, 2005	Arkansas
May 5, 2005	Georgia
June 1, 2005	North Dakota
June 28, 2005	Delaware
July 1, 2005	Florida
July 1, 2005	Tennessee
July 24, 2005	Washington
September 1, 2005	Texas
December 1, 2005	North Carolina
December 8, 2005	New York
January 1, 2006	Connecticut
January 1, 2006	Louisiana
January 1, 2006	Minnesota
January 1, 2006	Nevada
January 1, 2006	New Jersey
January 31, 2006	Maine
February 17, 2006	Ohio
March 1, 2006	Montana

Effective Date	*State*
March 1, 2006	Rhode Island
March 31, 2006	Wisconsin
June 8, 2006	Oklahoma
June 22, 2006	Pennsylvania
June 27, 2006	Illinois
July 1, 2006	Idaho
July 1, 2006	Indiana
July 13, 2006	Nebraska
September 1, 2006	Colorado
December 31, 2006	Arizona
January 1, 2007	Hawaii
January 1, 2007	Kansas
January 1, 2007	New Hampshire
January 1, 2007	Utah
January 1, 2007	Vermont
June 29, 2007	Michigan
July 1, 2007	District of Columbia
October 1, 2007	Oregon
January 1, 2008	Maryland
February 3, 2008	Massachusetts

Looking at the list in Table 3-1, you might say "Wait! What if I live in South Dakota, which hasn't passed a security-breach law yet?" Fear not — you're still very likely to get a notification if your biometric information is exposed by a security breach. As it turns out, most of the businesses to which you may have given such information also do business in one of the thirty-nine states that *do* have security-breach laws. You'll be notified, all right. That's because trying to sort out the people who reside in states that passed such a law from those who live where no such law exists is a real pain — and you can't be sure another such law didn't get passed in the "uncovered" states while you weren't looking. If a security breach compromises 10,001 records and only one of those people lives in South Dakota, it's not considered good customer service to tell the *other* 10,000 people, "We're very sorry but your private information has been compromised by a hacker. Please don't tell anyone from South Dakota — especially if you're a member of the press."

The California law and most of the ones that follow it can be summarized pretty simply: If personal information (which we define in a moment) is exposed by a breach in your security, you must notify all the persons whose information was exposed. There is some detail about how to accomplish the notification and exceptions for ongoing legal investigations but that's the law in a nutshell. Most states use the same definition of *personal information* that California does — which does not explicitly name biometric data — as follows:

> *An individual's first name or first initial and last name in combination with any one or more of the following data elements, when either the name or the data elements are not encrypted.*

✔ *Social Security number*

✔ *Driver's license number or California ID card number*

✔ *Account number, credit or debit card number, in combination with any required security code, access code, or password (e.g., a PIN) that would permit access to an individual's financial account.*

Although this definition does not explicitly call out biometric information, it could easily be argued that fingerprints or other biometric samples constitute a "required security code" or "access code" to any reasonable jury. Our bet is that few companies would be willing to risk the public-relations problems that could arise from not informing customers that their biometric information had been stolen — and then having that information announced by the news media instead. Kudos to Florida, Nebraska, Nevada, North Carolina, and Wisconsin — all were forward-thinking enough to explicitly describe biometric data as personal information in their breach-disclosure laws.

Health care to the rescue

An idea that has not been tested in court yet is that, in many cases, biometric information could be considered health-care-related, called Electronic Patient Health Information (EPHI) according to HIPAA — the U.S. law that defines how EPHI must be protected in computers and networks. Since changes to some of the biometric data for an individual can only be explained by certain changes to their health status, employers must be very careful about how they handle iris, vein, retina, and other biometric measures. They have to be certain they don't run afoul of our nation's sensitivity to the privacy of health-care information. Current laws and regulations don't come into play directly, but the U.S. as a nation has shown an intolerance to exposing private data regarding the health of citizens — to a degree that should give pause to organizations collecting this information as they decide how to protect it.

Biometrics and European Law

Europe — and more specifically the European Union (EU) — is very proud of the special care it takes with the privacy of its citizens' personal information. Unlike the U.S., where the fundamental right to privacy is somewhat obliquely referred to in the Constitution, the EU Constitution has a comprehensive privacy statute:

Directive 95/46/EC. Also known as the Data Protection Directive, this statute deals directly with the handling of personal information — including biometric information. (In defense of the U.S. Constitution, computers and biometric information weren't popular topics of discussion in 1787 when the Constitution was written and ratified into law.) The full text of the EU constitution can be found at the official EU Web site:

```
http://europa.eu/scadplus/constitution/index_en.htm
```

Article II-68 of the EU constitution titled "Protection of Personal Data" is refreshingly short and clear. The entire text of II-68 consists of the three assertions:

> *1. Everyone has the right to the protection of personal data concerning him or her.*
>
> *2. Such data must be processed fairly for specified purposes and on the basis of the consent of the person concerned or some other legitimate basis laid down by law. Everyone has the right of access to data which has been collected concerning him or her, and the right to have it rectified.*
>
> *3. Compliance with these rules shall be subject to control by an independent authority*

With these guiding principles, the Data Protection Directive is applied to the "processing of personal data," which is defined as "any information relating to an identified or identifiable natural person." The guiding directives from the directive concern the following topics:

- ✔ Principles relating to data quality
- ✔ Criteria for making data processing legitimate
- ✔ Special categories for processing
- ✔ Information to be given to the data subject
- ✔ The data subject's right of access to the data
- ✔ Exemptions and Restrictions
- ✔ Data subject's right to object
- ✔ Confidentiality and security of processing
- ✔ Notification

The actual directive includes pages and pages of detail regarding each of these topics. It boils down to the idea that an EU member citizen's data must be carefully collected (if collected at all) — and the information collected must be enough to accomplish the required information-processing task *and no more*. In addition, the information must be handled exactly as described to the citizen, who must be allowed access to information collected on him or her, and the information must be carefully protected from disclosure at all times.

The effect of these privacy laws on biometrics in the EU is interesting from a social point of view. EU member nations' citizens are generally more comfortable with the collection of extensive biometric information by their governments than U.S. citizens are. That's because the collection and use of such information is so much more carefully controlled than it is in the U.S. Nearly all member nations have biometrics-based identification built into their passports or have that capability and will soon be using it.

Since there are EU laws regarding how personal information can and will be shared between EU nations, port-of-entry security between member nations can be far more relaxed than it is in countries that don't have laws that restrict the handling of personal information such as biometrics. This also allows companies to collect information from citizens of other EU nations without too much additional scrutiny of information-handling procedures. This is in stark contrast with the hoops that American companies must jump though to handle EU personal information; U.S. law is not nearly restrictive enough to satisfy European sensibilities. Even the US-VISIT program, which collects fingerprints from all people entering the U.S., was subject to scrutiny by the European Commission as a potential violation of Data Protection Directive.

The EU laws closely resemble proposed "Guidelines on the Protection of Privacy and Transborder Flows of Personal Data" from the Organisation for Economic Co-operation and Development (OECD), an international organization with thirty member countries, including the United States. The OECD guidelines steer the privacy and information sharing policies of nearly all member nations except the United States.

Biometrics and Laws in Other Countries

The OECD guidelines use the same basic guiding principles already listed in the section on the EU Data Protection Directive — but guidelines issued for the member countries don't carry the force of law or Constitutional authority. The force that the guidelines do have, even outside of the member countries, is that for the purposes of border protection, anti-terrorism, and law enforcement, biometric information must be shared among the nations of the world — and the OECD guidelines are designed to be restrictive enough to satisfy privacy concerns (at least as well as can be imagined in such circumstances). Even the U.S. has endorsed the guidelines, but as yet has done little or nothing to comply.

In general, democratic governments are doing a balancing act: the needs of society to control borders, catch criminals, and prevent identity-related crimes (such as identify theft) against the individuals' rights to privacy. Nondemocratic governments, on the other hand, have been quick to adopt biometric identification of citizens — in many cases, for the purpose of tracking their movements and activities.

Currently, there are no laws that describe the idea of what we call *tainted biometric data*. Biometric information collected from a populace who have no choice in the matter should make us think twice about the use of such data in our efforts to make the world a better and safer place. We would consider any biometric information collected against the will of the individual or coerced from an individual to be tainted in this respect. Other ways that biometric data might be tainted would include collection without the subject's knowledge or gained from sources like health-care organizations that collected the information for non-biometric uses.

Ethics Issues with Biometrics

The use of information to identify or authenticate persons gives rise to ethical considerations surrounding improper use of the identification and authentication processes themselves, but in the case of biometric information, some of the measures are so personal that the biometric measure itself can raise ethical concerns.

Tracking you down

Any measure that can be used to identify a person might also be used to track their movements and activities in ways completely unrelated to the task for which the identifying information was originally collected. In the case of biometric data, there is no way to decouple the identifying information from the person — resulting in a really sticky situation: Every place where that information is captured can be correlated with every *other* time and place that information has been presented, forming a complete picture of the individual's activities, tastes, financial position, associates, affiliations, and whereabouts. Although that may be desirable when talking about criminal activity, it's an extreme violation of personal privacy for normal citizens going about their noncriminal lives.

Some biometric measures — such as gait, iris, facial imaging, and voiceprint — cannot easily be concealed in public places, and might allow entities in possession of this kind of biometric information to know far more about the subject than would be possible otherwise. Indeed, many of these biometric measurements can be captured without the subject's knowledge or consent. In the absence of good laws preventing such actions, we may soon see gait-based biometrics used to follow your progress though the mall — and making suggestions to store clerks about what products to offer you, based on your earlier purchases or store visits. Then again, it could be relatively easy to wear a "disguise" in the form of a heel lift in one shoe that will change the way you walk, and allow you to "disappear" completely!

Controlling how your biometrics are used

Good privacy laws make sure that biometric information collected from an individual *with* her consent may be used only for the precise purpose for which it was provided, and no more. Privacy advocates who are against the use of biometric identification have the legitimate fear that once the information is collected, it cannot be withdrawn; that biometric identification is forever associated with the subject, even if the information eventually falls into the wrong hands or policies change to a point at which the subject would not have willingly given the biometric data had those policies been in place at the time.

Currently, fingerprints captured in the US VISIT program are used only to control entry by undesirable persons, and to capture exit time and place for non-U.S. travelers. Another terrorist attack on U.S. soil, however, might change how we look at that data entirely — and cause us to start profiling travel behavior using biometric entry and exit data. Not only would that potentially violate our agreements with the countries where these travelers originated, it would violate the travelers' understanding of how that information was to be used when they supplied it.

Ignoring irrelevant data

Biometric information by its nature may also capture information regarding race, ethnicity, gender, and other potentially prejudicial physical characteristics. From long experience, we know that some humans have a hard time ignoring prejudicial characteristics if they become known. The potential for irrelevant characteristics becoming available in facial-recognition systems that capture and save the entire image — and not just the metrics — are obvious, but other biometric measures also have this potential as well. At the very least, some physical differences such as missing limbs, weight, height, or eye color will be obvious with several biometric identification systems.

The irrelevant data should be discarded, to whatever degree is possible, to prevent prejudicial behavior on the part of people with access to the biometric data; in addition, regulations must be put in place to protect subjects when it's impossible to eliminate such information. For an example of a primitive identification that elicits prejudicial response and even fear, consider surnames — and the behavior that ethnically identifiable names can produce.

Understanding failure

Failure to authenticate or to be identified by a biometric system can be embarrassing and potentially far too revealing. Reasons for failed authentication can include

- ✔ Weight loss or gain
- ✔ Hair loss
- ✔ Medical conditions
- ✔ Body changes due to trauma
- ✔ Change in personal hygiene
- ✔ Psychological problems (behavior-based)

Each of these potential reasons for a false rejection can have the potential to cause the subject loss of self-esteem, loss of personal privacy, or just embarrassment — none of which should not be a potential outcome of logging in to the office system in the morning.

 People implementing biometric systems need to be aware that failures to authenticate can be based on highly personal, and sometimes embarrassing, factors that have nothing to do with the technology and everything to do with users' private lives. Sensitivity to this issue and good training for people operating the systems can help prepare you for somewhat awkward situations with your users. We cover similar user support issues in more depth in Chapter 9.

Managing concerns

There are very few ethical concerns related to biometrics that are not countered by careful consideration of users of the system and their potential concerns. As long as the use of such systems is voluntary — and the information collected is used only in the precise way described to the subject at collection time — biometrics can be used ethically and responsibly.

The U.S. is sorely in need of laws that address privacy more specifically, if not a constitutional amendment that addresses a right to privacy. In the absence of such constitutional clarity, however, interpretations of legal precedent and current laws will leave citizens wondering what will happen to the biometric information gathered by the government, commercial applications, and (for example) shopping mall security — and how that information may be used. In the meantime, it's important for users to understand who is getting their hands on such information, and how — whether? — they have agreed to protect it and use it.

Part II
Types of Biometrics

The 5th Wave   By Rich Tennant

Here he comes. Remember, when he opens the cage, go for his eyes.

"Come on in. I left my new voice recognition system on so you could try it."

In this part . . .

In what ways can you measure a man (or a woman)? With biometrics, those ways range from fingerprints, hand scans, and hand-vein scans to retina and iris scans and facial recognition. Also there's handwriting, voice recognition, and keyboarding. Don't forget gait (how you walk) and facial thermograph (no need to blush about that). Then there are more exotic methods such as odor, DNA, and brainwave — and even a few more in this section that we don't want to give away here.

Chapter 4

Fingerprint and Hand Biometrics

• •

• •

*I*n this chapter, we will get a look at fingerprint and hand-based biometrics to understand their strengths and limitations in practical use. Hands and fingers are pretty obvious choices for biometric identification since they have many characteristics that are unique to the person they're attached to, and we're used to poking them into places where we wouldn't want to put our eyeballs, ears, or other body parts that have sufficient uniqueness. This propensity to use our hands in all kinds of ways also presents unique challenges for biometrics that we will discuss here as well.

Any biometric measure has what we call a *biometric basis,* which is the basic concept that allows us to use that measure to uniquely identify a person. For each kind of biometrics we discuss, we briefly explain the idea behind the particular biometric measurement as the biometric basis.

Fingerprints

The inner surfaces of our hands and feet are covered with ridges and furrows. The formation of these ridges happens in our mothers' wombs, only partially guided by our specific genetics. Thus, while identical twins tend to have similar characteristics, the details of each area — the minutiae — are unique to the individual. With such reliable uniqueness conveniently packaged at our very fingertips, it's not hard to understand why fingerprints are the most popular and widely used form of biometric identification.

Understanding the biometric basis for fingerprints

An image of the pattern of ridges and furrows that cover our hands, feet, and fingers can be optically captured at a point in time when we're certain of the identity of the provider, and then later compared to a new image to authenticate that user. Although fingerprints change size as we grow, the structure of the ridges and furrows doesn't change at all over time, except for essentially mechanical alterations such as cuts or scars. Figure 4-1 shows a close-up of fingerprints.

Figure 4-1:
Close-up of
fingerprints.

Contemplating practical considerations

We use our hands a lot, so the possibility that our prints become marred by damage, temporary or otherwise, is much greater than that of damage to, say, our eyes or facial structure. Physical damage to your fingertip means it doesn't look the same as when you captured your print for identification — and the system may reject your identification if it can't exactly match one or more of the unique characteristics.

Although — in theory, anyway — we could use any portion of our feet and hands that have identifying characteristics, in practice the tips of each finger or thumb are the easiest to position for imaging. They're so easy, in fact, that each of us leaves perfectly legible copies of our fingerprints on objects we touch all the time — as many criminals have learned the hard way. While this makes finger- and handprints very convenient as a biometric identification, it also makes finger- and handprints among the best-documented (and most generally available) biometric identifiers we have.

Misappropriation of fingerprint information isn't commonplace (yet). In one well-published case, however, the German Interior Minister's fingerprint was acquired from a drinking glass at a public event, and then published in a magazine, along with the means to use the print to fool some fingerprint readers. As far as we know, nobody has successfully used the published print to impersonate the minister, but it's not beyond the realm of possibility. To read more about this prank, take a look at this article located online here:

```
www.theregister.co.uk/2008/03/30/german_interior_minister_fingerprint_
                appropriated/
```

Fingerprint, and by extension palm-print, readers come in essentially three forms:

- ✔ **Optical:** These work much like a regular image scanner, where a light source is used to illuminate the surface of the scanner area and a charge-coupled device array collects an image of the illuminated surface.

- ✔ **Thermoelectric:** Thermoelectric scanners use substances that electrical properties are influenced by localized heat sources (like your finger) and read the electrical variances in the surface to acquire an image of the fingerprint.

- ✔ **Ultrasound:** Ultrasound imaging of fingerprints bounces very high frequency sound waves off the three-dimensional structures of your fingerprint and records the 3D model acquired.

Fingerprint identification is famously challenging when the prints are collected from wine glasses, door handles, and various blunt instruments. In a biometric identification scheme, however, you're working in a controlled environment — and can use that greater control to get more reliable results. In other words, if we are consciously *trying* to leave a good fingerprint when registering on a bio-metric system, chances are we'll leave a better one than we might on the handle of that big kitchen knife when we have something else in mind.

Scanners for fingerprint biometrics come in a number of shapes and sizes, with two basic modes of operation

- ✔ **It scans you:** Your finger or fingers remain motionless while a scanning device moves across your print to capture it.

- ✔ **You scan it:** You move your finger across a stationary scanning device while it captures your print.

The second method — moving your finger across a stationary scanning device — is only practical because fingerprints are pretty simple to under-stand, and not much open interpretation in positioning your finger for scan-ning. This is not true for some other prints — in particular, palm scanning (more about that in the next section). Figure 4-2 shows a fingerprint scanning device in use today.

Say AAA

You'll often hear security wonks talk about *Triple A* as if normal people would have any idea what that means. The As in AAA stand for

✔ **Authentication:** Who are you?

✔ **Authorization:** What are you allowed to do?

✔ **Accounting:** What is actually being done in the name of the authenticated, authorized person?

Biometrics is almost always about who you are, so we're mostly dealing with authentication in this book. There are, however, some kinds of biometrics that folks are starting to look at for specific authorization and access applications, such as "Based on pulse, pupil dilation, and facial blood flow, this person is too angry to be allowed to send this e-mail. We'll let him calm down a bit first." (Perhaps there's a solution to road rage here, similar to built-in breath tests in cars. . . .)

For more about Triple A principles, see Chapter 2.

Figure 4-2:
Typical fingerprint applications include computers, doors, and safes.

Where you will see fingerprint biometrics

Fingerprint biometrics serve as the workhorse of biometric-based identification. Typically, you will see fingerprint biometrics on door locks, laptops, keyboards, and more formally at points of entry. Three factors have largely influenced the widespread use of this tool:

✔ The technology required to do simple fingerprint scanning is relatively cheap.

✔ The analysis required to determine a match (or lack of one) doesn't require a lot of computer resources.

✔ We humans seem to be more comfortable sticking our fingers — rather than our eyeballs or other body parts — into the ID device.

On the other hand, simple fingerprint systems have been shown (on some occasions) to be relatively easy to compromise by using fake fingers with real fingerprints. As a result, fingerprints are typically used for low-to-medium-security applications (such as logging in to personal workstations) — with notable exceptions, such as the US VISIT program (which we discuss in Chapter 2). Chances are a more sophisticated biometric method will be used for a high-value asset or system.

Fingerprints are sometimes used in higher-security settings when other factors are available to corroborate their use. For example, the fingerprints contained in the electronic memory of certain passports, or on the new Hong Kong Identity Card, are always used in the presence of a security person, such that using a fake finger would likely be obvious, unless you are a spy with amazing skin-thin fake-print gloves. In that case, you could also have altered your retinas and hand-vein structure as well. (In that case, you'd better check: You may be a fictional character in a movie.)

Palm Scan

The same kind of ridges and furrows that make our fingertips unique also cover the inside of our hands (and feet, for that matter). Our palms also have another feature that is unique when studied in detail: the *flexion creases*. In addition to the conspicuous (typically three) flexion creases in each hand — formed at around $2\frac{1}{2}$ to 3 months after conception — a number of finer lines with less regular patterns are formed at around 4 months. Combinations of these characteristics are what we refer to when talking about palm scanners and palm prints in biometric identification. Figure 4-3 shows what we are talking about here.

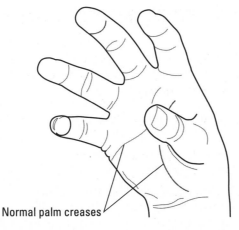

Figure 4-3:
Palm flexion
creases
are another
popular
biometric
identifier.

Normal palm creases

Understanding the biometric basis for palm scanning

Like fingerprints, the ridges and furrows, as well as the flexion and other creases in our hands, are formed before we take our first breath of air, and don't appreciably change throughout our lives. While our wrists and palms are not quite as flexible and easy to position for imaging as fingertips, they are still, um, handy for placing on scanners of various shapes and sizes to collect an image.

A full image of an entire hand, including the palm and fingerprints, provides a large number of minutiae for comparison — more than for fingerprints alone — resulting in accurate matches with few false negatives or positives. Palm scanning is typically more restrictive about positioning the hand with respect to the scanner, which also tends to provide better accuracy, since the palm is more likely to be in the same position each time.

The larger number of points collected in a hand scan means that more changes, like a paper cut, can occur before the comparison to the original is too changed to recognize, which still providing a high level of assurance that the palm presented is the same as the reference sample.

Contemplating practical considerations

While palm scanning can be very accurate and the larger scanning area more forgiving of changes to the image, palm scanners tend to be more expensive than fingerprint scanners for a couple of reasons:

- **Palm scanners must be more substantial to accommodate everyone's hand and the greater weight of the palm.** Something you rest your finger on can be much lighter, smaller, and less expensive than something intended to scan your palm. Designers have to take into account the fact that few people would lean on a finger while authenticating themselves, but it's second nature to lean on something your palm is resting on.

- **Palms are much larger than fingers, so the imaging surface must be much larger as well.** In electronic imaging, the size and quality of the imaging surface is a large part of the cost of a device. The palm surface isn't naturally a flat, or nearly flat, surface: it's curved. Getting a really good image of the palm requires an imaging surface that's curved to capture the print in a relaxed state and not stretched. The mechanics of scanning from a curved surface are much more complex and expensive than those that scan from a flat surface.

A factor in favor of palm scanning is that while we do tend to leave our fingerprints on most of the smooth surfaces we touch, we leave palm prints much less, and those prints rarely include the flexion crease details. FBI statistics show that about 20 percent of the prints collected at crime scenes are palm- or handprints compared to the other 80 percent that are from fingertips. That means palm prints are far less likely than fingerprints to be collected by some hacker with the intent to misuse them.

Where you will see palm scan biometrics

Palm scanning is losing ground to hand *vein* scanning, which we talk about next, but it can still be found where security is important — in places such as IT co-location facilities and high security building access. Palm scanners are good at dealing with changes over time when the authenticating customers are a large group, and the time between authentications may be months or years. Because palm scanning is forgiving of minor changes to the appearance of the palm, it is quite suitable for IT hosting providers, which may only be visited a few times each year. In the intervening month you may suffer cuts or abrasions to your palm or even scars, but there should still be enough distinguishing characteristics to authenticate on the first try. If a more picky method, such as fingerprint scanning, were employed in this example, the hosting provider would have to deal with frequently re-authenticating customers to allow access to their facilities.

Palm scanning and palm printing are now in frequent use in law enforcement to provide a more complete analysis of crime scenes. About 20 percent of the prints taken from crime scenes are palm prints.

Hand Veins

Unlike other hand-based biometrics, hand-*vein* biometrics focus not on features that are unique to hands and feet, but on veins, which occur throughout the body. Although theoretically possible to use the unique structure of veins in other parts of the body, the hand's unparalleled utility for placement on scanners is an important reason that hand-vein scanning — and not, say, elbow-vein scanning — is gaining ground as a very secure, relatively nonintrusive method for establishing credentials. Another important factor will be familiar if you played with a flashlight as a child, shining it through your fingers to see the shadowy bones and vessels inside. The hand, fingers, and wrist are all thin enough to let light through, illuminating and measuring internal structures — especially if we use some interesting tricks with colored light.

Understanding the biometric basis for hand veins

The veins in your body are somewhat similar to the veins in any other person's body at the macro level, but there are enough differences to make the false rejection rates (FRR) and false acceptance rates (FAR) better than any other hand-based technology. Because finger and palm prints are used in courts of law, sometimes with life and death hanging in the balance, a number of studies have been conducted to show just how unique those biometric indicators are. Hand veins are never used in such circumstances, and the relative uniqueness is less well documented. So far, private studies and experience indicates that vein patterns are at least as unique as finger and palm prints, and the images lend themselves to analysis better than any of the surface methods, making hand veins a generally more reliable method.

An image of the vein structure is taken by shining an infrared, or near infrared light, into the hand, which makes the red blood in veins appear black, while the surrounding flesh appears white. This results in a high contrast image of the vein structure suitable for comparison in subsequent authentications. Figure 4-4 shows an example of a hand-vein scanner.

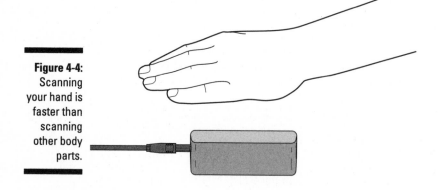

Figure 4-4:
Scanning your hand is faster than scanning other body parts.

It's alive!

Although this isn't the most pleasant of subjects, the idea of knowing whether a biometric sample is alive when it's presented is important — and in a few documented cases, it's a practical consideration. In at least one recorded case, a Malaysian businessman lost a finger to car thieves who wanted to be able to start the car without him later. A system that can see whether the subject it's scanning is alive protects both the asset *and* the personnel by making that gruesome gambit useless.

Contemplating practical considerations

Hand-vein scanning has nearly all the advantages of other hand-based biometrics, and eliminates nearly all the disadvantages.

Since the scanning technique ignores the surface of the hand, and only captures internal vein structures, the problem with minor cuts, abrasions, and new scars is pretty much eliminated. Nothing short of catastrophic injury to the hand will change the image of vein structure enough to produce a false rejection.

As with palm scanning, positioning is relatively easy and tends to be consistent from one scan to the next — and the larger number of data points (compared to fingerprint scans) means that even the unfortunate loss of a finger (for example) would not change enough of what the scanner sees to invalidate the authentication, and the user would be properly recognized.

One of our complaints against finger- and palm prints is the use of something that we leave on *everything we touch* as a key to authenticating people. The unique pattern of the veins in our hands, on the other hand, will not be left behind on wine glasses and candlesticks as we go about our day. Even if someone did end up with a good image of your hand veins, nobody has demonstrated a simple way to produce a fake hand that will mimic the pattern well enough to fool a scanner. (Yet.)

On the ghoulish side of things, a severed finger or hand will typically fool a fingerprint or palm-print scanner (so we're told) but that same finger or hand will not produce a good vein image, so it can't be used to gain access.

Where you will see hand vein biometrics

Until recently, hand-vein scanners were seen only in high-security applications as physical-access authentication devices. As with palm scanning, hand-vein scanning works well for facilities access. Vein scanning's ability to ignore small changes to the skin and its great FRR and FAR statistics make it a great solution when you have a large number of users to track, but not a lot of access to those users for updating their biometric information. You will see hand-vein scanners in high security facilities for entry points and secure room access.

More recently, Fujitsu has released a mouse with an embedded hand-vein scanner that brings a very high level of surety to biometric identification at the workstation level. Infrared imaging isn't technologically difficult, so you should expect to see various kinds of vein scanning built into smaller devices, such as mice, to realize the many advantages of vein scanning over the most popular current portable technology, fingerprinting.

Sonar/Ultrasonic

It was a little tough deciding whether we should put sonar and ultrasonic technology into Chapter 11, where we discuss biometrics' future, or here because presently it deals primarily with hands. We've chosen to compromise and discuss the current work in hand biometrics (because it's seeing some use today), and leave the other sound-based imaging concepts (which are still just ideas at this point) for Chapter 11.

In general, this technology relies on the fact that it's possible to use sound waves to map a three-dimensional space with a fairly high degree of accuracy. Possibly the best-known use of this approach is medical ultrasonic imaging, used to get pictures of the inside of our bodies for diagnostic purposes. Since ultrasonic imaging doesn't involve potentially harmful radiation (as does the use of X-rays), we use it to look at fetuses in real-time video. If my friends and family are any indication, there's a whole generation of children coming whose first baby pictures are taken this way (and shared on the Internet).

Understanding the biometric basis for sonar and ultrasonic biometrics

This chapter follows the progression of scanning, from the simple surface images of fingerprints through progressively greater detail of the structures in and around the hand. Ultrasonic measurement can detail the internal structures of a hand — including veins, muscles, bones, and even such extreme detail as the 3D structure of the fingerprint *from the inside*. These measurements can be represented as two-dimensional images, or (using a technique called ultrasonic holography) store all the 3D information for comparison.

As you might expect, the wealth of information about the structures in a hand that are collected in this manner are unique to each individual — and impossible to duplicate with current technology. Ultrasonic imaging is able to detect blood flow, for proof that the hand presented is alive (cloning fingerprints may be easy, but cloning a living hand is considerably more difficult, we think), and might even be able to tell how much stress someone is under by looking at heart rate and blood-vessel dilation (indicators of response to duress).

Contemplating practical considerations

Most current work in ultrasonic biometrics and the systems that demonstrate this approach have focused on the simplest use of the technology: capturing a 3D image of fingerprints. The more sophisticated kinds of imaging and analysis use more information from the hand or other body parts — and they're just around the corner, almost ready for practical use.

The patents filed for these devices claim that they can be produced cheaply enough to incorporate into workstations and laptops, which would make this one of the few biometric scanning systems available at this level that can prove the sample it's scanning is alive.

Where you will see sonar and ultrasonic biometrics

Of all the hand- and finger-based biometric systems discussed so far, this is the one you are *least* likely to see in current use. Several companies and universities are currently testing this technology and figuring out where it might best be deployed, but not a lot of consumer-ready equipment is out where you might see it. Based on the advantages of this technology and the relatively low cost of ultrasonic *transducers* (the things that emit the sound waves and receive the echo), this technology could easily show up on laptops in the next couple of years.

Comparing Hand-Based Biometric Types

The following table provides a quick comparison of the hand-based biometric types that we discuss throughout this chapter.

	Strengths	Weaknesses	Cost	Counter-measures	Convenience
Fingerprint	Simple, cheap	Easily spoofed	Low	Live guards to watch the process	Good
Palm scan	Able to deal with small variations	Cost, large scanner	Medium	Live guards	Good
Hand vein	Accurate, proof of life	Cost, not yet mature	Medium to high	Largely unneeded	Excellent
Sonar	Accurate, able to deal with small variations	Cost, not yet mature	High	Unknown	Good

Chapter 5

Signature Biometrics

. . . .

. . . .

*T*his whole chapter focuses on the most commonly used biometric measure in formal contexts — signatures. Society has long accepted a handwritten signature as formal authentication for documents. The Talmud describes a process for signatures and witnessing signatures as early as the third century, and by 500 A.D. the Roman Empire was using the subscripto, a handwritten subscription to the contents of a document for wills, which was quickly adopted for other legal documents and shortened to be just a representation of the signer's name.

Since the early Roman uses of signatures there is an unbroken history of their use, formalized by England in 1677 in a law called "An Act for Prevention of Frauds and Perjuries" — which is the basis for accepting written signatures for document authentication today. We should also note that unlike any other form of biometric authentication, signatures also often convey *intent*. Signing a loan form doesn't just show that you were *there* and signed the document; it shows your intent to *agree* with the terms of the document. Even though (as this chapter shows) signature biometrics fall short technically when compared to many other biometric measures, the unique element of intent makes the biometrics of signatures a more popular technique than it would be otherwise.

Although signatures are widely accepted as a tool for authenticating that a specific person was present and signed a paper document, the current practical use is primarily limited to an examination of the signed paper document to verify that the signature matches signatures that are known to be from that individual, comparing the shapes of letters, the construction of letter shapes, and the pressure applied in creating the signature when that information is available.

Signatures in a biometric setting use all of that information, but instead of examining an already executed signature, a biometric signature authentication system has the opportunity to capture information *during the act of creating the signature* and to use accurate measurements of various other factors to collect a good *behavioral* biometric sample. Although all signatures are (technically speaking) examples of a behavioral biometric, this chapter concentrates on technology-driven signature biometrics.

Recognizing a Signature

For the last several hundred years, signatures have been recognized using the simplest possible method: visual matching. If the signature in question closely matched the appearance of a known sample, it would be judged to be from the same person as the known sample. As we became more sophisticated about signatures and dealing with forgeries, we started looking at nuances of the shaping, strokes, and pressure when possible. For example, an original signature (one that has not been reproduced by photocopying, faxing, or other means) retains details such as the varied pressure applied to the paper while the signer was creating it. Using sensitive instruments, we can actually measure the depth of the slight groove created as the pen moved across the paper. Depending on the paper, you can also measure the speed of the pen in various areas of the signature, depending on how the ink bleeds into the surrounding paper because slower strokes deposit more ink, which bleeds more. Strokes that leave the paper also tend to trail into a thin point rather than ending abruptly. The way this happens varies from person to person.

Many of these items are fodder for biometric signature recognition, since they can also be measured directly while the signature is collected. Wet ink signatures can also supply other factual aspects that signatures collected purely electronically cannot — for instance, ink composition and color, paper absorption and bleed, and the aging characteristics of some inks. Although these chemical and physical characteristics are interesting and useful for authenticating signatures from historical documents or paintings, they're not all that helpful in authenticating a credit-card transaction.

Understanding the biometric basis for signature recognition

The most basic form of signature recognition for biometrics is quite similar to what happens in fingerprint biometrics. We call this basic form an *image-only* biometric signature since we are not using anything about the *act* of signing — only the image of the signature itself — to authenticate the signer. The method collects various characteristics of the signature from a known signature sample (or multiple samples) and compares their characteristics to the sample presented for verification or authentication.

Full comparison of any two signatures will almost always fail, since it's nearly impossible for anyone to exactly duplicate a signature. Minor variances in paper, ink flow, pen weight, and muscle control will nearly always introduce variations between any two signatures, even from the same person.

To deal with these variations, signature comparison algorithms use other characteristics to help correctly identify authentic signatures. In later sections of this chapter, we explain how movement and pressure are used to help with this process; right now we are primarily concerned with the physical image of the signature.

Although any two signatures from the same person may not be identical, the relationships between the letters, the relative sizes of loops and character spacing should match. For example, if your lowercase letter *t* is exactly three-quarters the height of your letter *l*, that ratio will generally remain constant across all your signatures. Are the two letter *t*s in *Littleton* crossed with a single stroke or with two? From the left or the right? Do you pause after the second *t* for the cross-stroke, or wait until the end? Do you bother to dot the *i* or leave it undotted? Is your *i* dot a speck, a line, a circle, or more comma-shaped? Each of these characteristics, though it may differ somewhat from the sample, will be repeated in ways that can be compared to the original and authenticated.

Letter shape is also an obvious place to look for uniformity and biometric uniqueness, but in many cases it's behaviorally less distinct (and physically more subtle) than other gross characteristics of the physical form of the signature.

Contemplating practical considerations

Any signature is a tough biometric to do well when all you capture is an electronic image of the handwritten signature (the technique we examine in this section). Two measurements that are important for signature biometrics are Failure To Enroll, or FTE (the rate at which a biometric system is unable to enroll new users), and Failure To Acquire, or FTA (the rate at which a biometric system is unable to read a biometric sample from a user). Variations between signatures from the same person can result in much higher FTE and FTA rates than variations in most other biometrics — which just means the system might require several known samples in order to enroll an individual. In some cases (including one or both of the authors), repeated signatures are so variant that it can be impossible to enroll in a signature-based system at all. Although this may not seem such a big problem, having valid signatures questioned too often can be annoying.

Electronic signature pads (see Figure 5-1) are easily available and relatively inexpensive if you are only collecting an image. These devices are really just very small digitizer pads, so the technology has been around for quite a while and other uses have brought the prices down. They are in everyday use in thousands of retail locations in the United States and elsewhere, where customers swipe their credit cards and sign their names with a special stylus that reproduces their signatures digitally on a small display.

Figure 5-1:
An electronic signature pad.

In some cases, authentication can mostly ignore the whole step of comparing a new signature to a known sample. For example, signatures captured without *enrollment* (the collection of a known biometric sample, positively associated with a person in the database) at a point-of-sale system are not compared to anything unless (or until) the transaction is questioned. The assumption in these cases is that a new signature sample can be collected from the purported signer and compared after the fact. Since the number of disputed or repudiated transactions is relatively small, these systems have no need to incur the increased overhead of trying to match a signature to an enrolled identity in real-time.

Since the relative capture cost for signatures is small compared to other biometrics forms like iris or retinal imaging, and the concept of using signatures to verify and authenticate is generally familiar, people are tempted to use signature biometrics in places where it might not be the best technology for the job. Even though capture of the signature is inexpensive, the kind of analysis required to do a good job of comparing signature images to known samples is pretty involved, and it requires more computer processing power than you will find in most embedded systems. This is especially true when the required authentication must happen in real time while people wait.

Given that we use signatures to authenticate everything from real estate sales to international treaties; it may sound odd to say they shouldn't be used for serious secure biometric authentication — but let us explain. When signatures are used on important documents — which really must be authentic — they are generally witnessed by a disinterested third party — in some cases, a notary public whose primary function is to authenticate signed documents by watching someone with a proven identity sign a document — and then using a serialized seal to record that fact. In the case of international treaties, there are typically lots of in-person witnesses, and thousands or even millions of people watching on television. In the absence of witnesses or other corroborating proof, image-only biometric signatures are not appropriate for high security or high stakes use.

As one last practical consideration of image-only-based signature biometrics, it's one of the few forms of biometrics that can be considered discriminatory in and of itself. A handwritten signature requires some level of literacy; although an X or other mark can sometimes be used on paper documents, an X doesn't contain enough unique information to use for a signature biometric. Additionally, a number of ailments or age-related conditions cause fine motor control to degrade — making signatures illegible and unusable for biometric authentication.

Where you will see image-only signature biometrics

Image-only signature biometrics is a rarely applied method. That's due (in part) to the practical limitations described in earlier sections of this chapter. In some low-risk applications, however you will definitely see image-only signature biometrics used.

Electronic signatures are not digital signatures . . .

. . . and neither one really implies a biometric sample. A fairly wide range of *electronic signatures* is possible; it includes any kind of electronic sound, symbol, or process is affixed to a document or record — by someone with sole control of that identifier — with the intent to sign the document. *Digital signatures*, on the other hand, are a *subset* of electronic signatures that uses a cryptographic method to ensure message integrity and authenticity. (Knowing the difference between these two signature types will impress everyone who already knows the difference.)

It wasn't me — I was nowhere near there

Many years of working in information technology have shown both your authors that apparently nobody was anywhere near the site of any IT disasters when they occurred. Careful examination of the logs and video surveillance often shows that not to be the case (so we get to have that conversation with *someone* about keeping soft drinks out of the computer room). In the area of authentication, we'd love to be able to do the same thing in a more formal way — to accomplish what we call *non-repudiation*, essentially proving that the repudiation of a transaction was (ahem) incorrect, and somebody really *was* there. Although biometric measures are generally good at non-repudiation (after all, they're part of who you are), weak biometrics — such as image-only signatures — can't prove much if someone claims the signature is a forgery.

If you have received a package from a certain brown-uniformed delivery service in the last several years, you have most likely signed a small electronic pad with a stylus to acknowledge receipt. If you have ever had reason to question the delivery of such an item, you have seen the pixilated low quality image capture of something that could be either a signature, or an antelope in a polyester pants suit. Even with the poor capture quality, this generally works for the delivery service because they also have a witness to the act (the delivery person) and more carefully recorded details like the exact time and location where the signature was taken.

Image-only signature biometrics really make sense only when other factors are present to validate the signature; there's only limited biometric information available with just the image, and in many cases the quality of the digitized image is poor. Expect to see this technology in places such as point-of-sale systems where there is a human witness to the transaction — and in places where the intent is to show that a human being was present, but repudiation of the transaction is not a big issue.

Stylus Movement Dynamics (X and Y Directions)

To make up for the weaknesses in capturing only the image of a signature, two-dimensional stylus movement dynamics captures more information in the form of *how* the stylus moved while the signature was created. Since the digitizer that's collecting the signature is connected to a computer, it's perfectly possible to record and analyze the way the signer moved the stylus across the surface to create the signature — as well as the resulting signature itself.

Understanding the biometric basis for stylus-movement dynamics

As with many other biometric measurements, more detail can yield more accuracy. Lack of detail in a biometric measurement can lead to problems:

- ✔ **Failure To Enroll (FTE):** Not enough detail is acquired to make the measurement unique to an individual, or the variance of the specific data acquired makes it unreliable or unacceptable.

- ✔ **Failure To Acquire (FTA):** Similar to failure to enroll, except that the enrollment process ignored the wide variance in the measured data and now the system rejects a valid user too often.

- ✔ **False acceptance:** To compensate for wide variance in the measured data, acceptance criteria are set too broadly — until (say) Mike's dog's paw print is considered similar enough to his signature to authenticate the pooch as Mike.

- ✔ **False rejection:** Even though there is variance in the measured data, the system does not compensate, so Mike has to sign his name six times to get it exactly the same as when he enrolled.

By capturing additional behavioral characteristics of the act of signing your name, it's possible to improve the odds for accurate authentication and identification for signature biometrics.

Speeding

One of the things you can watch in two dimensions (okay, *three* if you count time) is the speed at which the stylus moves as a signature is generated. As it turns out, the speed with which we form letters and parts of letters is even more unique to each of us than the form of the character itself — especially when you're performing the act of signing your name, since people do that so often.

A forger trying to forge a signature is interested in just two scenarios involving the speed of signing:

- ✔ **Forging the name in private with all the time in the world.** In this scenario, the forger doesn't have to even be able to look like they are signing normally. They can laboriously move the pen across the paper, even tracing on top of the sample they have to make sure to get the right look. Although we can't imagine this scenario in a biometric setting, it's the most obvious example of something a good stylus dynamics biometric would reject, even if the resulting signature were a perfect forgery. Since the speed through the act of signing wasn't even close to the same as the recorded biometric information, it would be rejected.

✔ **Forging the name live, in front of someone else.** This scenario seems as if it might hold more promise for a forger to beat a biometric system, since the bad guy has practiced the signature over and over to be able to look like he's signing normally when he creates the forgery. The would-be forger would be sorely disappointed by failure in this attempt — and the failure could happen because the thing he practiced was exactly matching the *form* of the signature, not the *movement* of the original signer. Subtleties in the speed of the stylus through the act of signing differ from one signer to the next; even watching a video of someone signing his or her name isn't much help; trying to match it exactly would be nearly impossible.

Stroking

Another way to apply two-dimensional movement dynamics to signatures that is less costly from a computer processing time than speed dynamics is stroke order and direction. If we all remembered our lessons from grade school and followed them to the um, letter, stroke order and direction would be exactly the same for all of us, since we would all be creating letters in exactly the same ways. Remember the illustrated diagrams with numbers and arrows for creating each cursive letter? As it turns out, like many other lessons from grade school many of us forgot or abandoned the letter of the law for lettering and now create many of our letters our own special unique way. Can any of you really execute a proper cursive capital Q anymore? See Figure 5-2 for a refresher.

Figure 5-2: Stroke drawing of a proper cursive capital letter Q.

Stroke order and direction by themselves will be largely similar to everyone else's order and direction. In conjunction with other measures, however, they can still add uniqueness to a biometric signature.

Contemplating practical considerations

Unlike pressure dynamics (which we cover next), two-dimensional (2D) dynamics are available from almost any digitizer you would use to capture signatures. That's because these devices track the movement of the stylus or pen *in real time* as it moves across the writing surface. Thus any system that currently captures image-only signatures could be pretty easily used to capture two-dimensional stylus-movement dynamics — without changing hardware. The hardware limitations will likely be in the form of computer capability at the point of collection or sampling; many of those devices are embedded in handhelds or point-of-sale devices with tiny little processors and not much memory.

To actually *use* the additional information, you'll have to enroll users into the system and capture the 2D dynamics for later comparison. This requirement matches the use and behavior of other biometric systems, but differs somewhat from the traditional (non-biometric) use of signatures; it also differs from some uses of image-only signature biometrics.

Unfortunately, two-dimensional stylus-movement dynamics depends on being able to track the movement across a signature that's very similar to the recorded one. People whose signature generally is too variable to enroll with image-only signature biometrics will also still have a problem with image and stylus dynamics combined.

Where you will see two-dimensional signature dynamics biometrics

Anyone who wants to use signature biometrics for a serious security application will have to enroll the users with multiple images of each signature — enough to provide for normal variance and good samples of the two-dimensional speed and stroke dynamics. Doing so eliminates the effectiveness of tracing copies of the proper signature and other forgery attempts. This requirement does, however, effectively eliminate the use of two-dimensional signature biometrics in many of the places that are currently appropriate for image-only signature biometrics. Enrollment with multiple signature images just isn't practical in most of those cases.

Since there are simpler, more accurate methods of authentication that use biometrics, 2D signature dynamics are likelier to crop up in places that already accept normal ink signatures but need further authentication and non-repudiation capabilities. Some examples would be loan documents (remember signing about a hundred times to buy a house?), authorization documents by government officials, and potentially some in-person banking transactions.

Stylus-Pressure Dynamics

Although we tend to think of the motion of signing our name as a purely two-dimensional movement, microscopic examination of a regular ink signature reveals that we apply varying amounts of pressure throughout the process. That variable pressure yields varying ink widths and indentations in the paper we're signing upon. You can see examples of this — without the microscope — at the beginning and end of strokes where the pen is moving prior to, or just after, contact with the paper. That's where an ink signature characteristically narrows to a point as the pressure from the pen is just beginning or trailing off.

Based on this fact, it's quite possible to also capture information about the pressure applied to the stylus in the vertical z-axis (if we assume the paper or pad itself is in the x-and y-axis). As a matter of fact, we are three-dimensional beings moving a pen or stylus around in a three-dimensional world, so there are quite a number of additional biometric measurements to consider when we start down this path.

Understanding stylus-pressure and other biometric signature dynamics

The pressure we apply during the process of signing is as behaviorally unique to each of us as the shape of the signature itself. When experts are examining forged signatures, *pressure dynamics* — as shown by indentations on the paper as well as ink width — are typically more important than the general look of the letters, since the look itself is far easier to copy.

With sensors in either the signing pad or the stylus itself, it's quite possible to measure the pressure exerted — while simultaneously capturing the image of the signature, along with speed and direction of travel for each element of the signer's name. In combination, all these elements yield an amazing amount of unique biometric detail regarding the act of signing one's name — which can be compared to the already-enrolled data to authenticate the signer.

While we're talking about ways to instrument the stylus for measurement, let's also consider what else of interest the pen or stylus can tell us. As an experiment, tape a two-foot-long stick to your pen and watch the end of it as you sign your name. Notice all of that wildly exaggerated motion? The extension shows you what is already happening at the non-business end of the pen when you sign something, just makes it more visible. Well, what if we put a three-dimensional accelerometer into the pen or stylus and recorded the motion both at the other end of the pen, but pen's motion through space when you lift it to dot and *i*, cross a *t*, or start the next word. Now we're capturing a lot of information — the image of the signature, the speed and direction of the stylus on the pad, the pressures exerted on the pad as the signature is applied, and the movement — in three dimensions — of all parts of the pen while the signer signs. Whew! There really can't be much else of biometric interest there, right?

Mostly right. What about the grip and grip pressure of the signer on the pen? The shape of the hand, as well as training and experience, will affect how you hold a pen — as well as the pressures you apply to the pen as you sign your name. To a degree, this starts back into the realm of physiologically based biometrics and away from behavioral, but all are quite valid things to look at as part of signature biometrics.

If these extreme levels of capture and analysis seem farfetched, you should note that prototype pens and signing systems have been tested with *all* these methods except grip and grip pressure — we added those just for fun. But you never know. . . .

Contemplating practical considerations

None of the technology required to capture z-axis information either in the form of pressure dynamics or even accelerometer-based pen movement is particularly expensive, but most of it is not yet standard equipment in most signature-capture devices, so hardware upgrades will be required in most cases to capture this additional information.

Although a full three-dimensional picture of pen movement through the act of signing adds to the richness of biometric information, it will be considered overkill in most cases. The level of information to be processed and analyzed in a simple image capture (versus high-resolution capture) of the complex movements of a pen in three dimensions is several orders of magnitude different — and only marginally increases the utility of signature biometrics for the applications currently using it.

Where you will see 3D signature dynamics

For the most part, this prospect echoes what we said about 2D signature dynamics, so we won't repeat that here. The additional information offered by 3D dynamics might be attractive to very high-security applications that require a traditional signature in addition to the biometric measurements. A government official signing a treaty or a judge signing court documents would be likely candidates for this approach.

Accelerate in all directions

Accelerometers are finding their way into home electronics everywhere. The most basic ones can sense acceleration on a single axis, so could sense that your iPhone is oriented vertically or horizontally since the earth pulls everything towards itself — at a rate equivalent to an acceleration of 9.8 meters per second per second. Some smart people figured out that the only time on earth that an accelerometer will sense no acceleration at all is in free-fall and put them into laptops to shut down the hard drives just before impact with the floor to save data.

There is a patent filed for capturing a "signing action" in midair using a camera, but a wireless pen device with accelerometers could accomplish something very similar and not require that you sign in the general direction of the video camera.

Because this sort of signature biometrics provides the highest degree of non-repudiation, good candidates for using it are those situations that currently use a normal signature to authenticate an action with potentially serious consequences, such as

✔ Signing out explosives for the testing range

✔ Signing the sonar contacts shift log

✔ Judge signing legal orders

In these cases, the signature is not being used to authorize an action, but to show what responsible party put their name to it.

Heisenberg Applied to Biometrics

Werner Heisenberg noted that when dealing with atomic particles, the very act of observing either their location or their momentum changes them; observing one aspect changes the other. Thus you can know where a particle is *or* how fast it's moving, but never both at the same time. The problem is that to know where the particle is, you have to bounce a photon off it, and when you bounce the photon off the particle, you changed the momentum of the atom by some uncertain amount.

The popular example of what is called Heisenberg's Uncertainty Principle has to do with the act of measuring the temperature of the water in a drinking glass. Considering that the thermometer itself will change the temperature of the water, one can see that the very act of measuring often changes the object being measured.

What does that have to do with signature biometrics, you ask? You learn to sign your name using pencils and pens and using paper sitting on a desk or flat writing surface. Thousands and thousands of practice signatures have happened in this environment — and that fact forms the very basis for signatures as a behavioral biometric.

In a perfect world, a signature biometric system would exactly duplicate these conditions and we would sign our name using a normal sized pen on regular paper while we watched the ink flow and form our regular signature. In practice, most current signature-based biometric systems are so far from this ideal that it's amazing they work at all. In effect, the act of observing your signature is changing the signature in unpredictable and uncertain ways. What we usually see in current signature-based biometric systems:

- Styluses that in no way resemble a pen
- Styluses that do not imitate the flowing action of a ballpoint or even the smooth flow of a fountain pen
- Digitizing surfaces that are too smooth, too rough, or just generally don't feel like paper
- Styluses with wires coming out the top
- Digitizing surfaces that are too small to sign in
- Digitizing surfaces that float in space with no place to rest your wrist
- Digitizing surfaces that do not allow you to see your signature as you create it or that show the signature on a screen time-lagged behind your actual signing

Behavioral biometrics really work only when the act of observing the behavior doesn't mess with the behavior. We're convinced that enrollment-based signature biometrics works only because the changes to behavior are consistent across all samples, and the captured data is never compared to a normal ink signature on paper. On the other hand, if a certain brown-uniformed delivery service ever needs to definitively prove that Mike signed for a package, he expects that the antelope in the polyester pantsuit will not help their cause.

Comparing Signature-Based Biometric Types

The following table provides a quick comparison of the signature-based biometric types that we discuss throughout this chapter.

Signatures	Strengths	Weaknesses	Cost	Counter-measures	Convenience
Image only	Simple, cheap	Easily forged	Low	Witnesses to watch the process	Excellent
Image with 2D acceleration	Nearly impossible to forge	Requires many comparison samples	Low	Protected connections to database	Excellent
Image with Stylus pressure dynamics	Follows traditional signature comparison standards	Somewhat more complicated signature pads	Low to medium	Protected connections to database	Excellent
Image with 3D movement dynamics	Excellent non-repudiation characteristics	Complicated capture hardware	Medium to high	Protected connections to database	Excellent

Chapter 6

Retina and Facial Biometrics

*W*hen allocating types of biometrics to specific chapters, it made sense to group together everything that had to do with some part of your head. Although many of the techniques are vastly different, there are advantages and disadvantages they have in common. In this chapter, we discuss retina, facial, and hair biometrics. (Just kidding about hair biometrics — we don't think anyone's doing that yet.)

All the various head-based biometrics require some sort of imaging that involves — you guessed it — your head. Even though many of us are comfortable sticking our fingers or hands into something to get a fingerprint, thermal scan, or something else that will distinguish us biometrically from others, many are somewhat less comfortable doing the same with their heads (after all, there are sensitive things in one's head like brains and eyeballs).

From a psychological perspective, we tend to associate features on the head with *identity* far more than we do other features such as fingerprints, hand geometry, or elbow prints. From an early age, we learn to associate facial characteristics with individuals and remember things like eye color, ear shapes, hair styles, and even nose shapes. On the other hand (so to speak), most people could not tell you whether their closest friends' fingerprints were arches or whorls.

These psychological components of how we think of and protect our heads all participate in how willing we are to use physiological biometrics that target features on our heads. It also colors how accurate and intrusive we expect these biometrics to be, especially with respect to facial recognition. Do you really want a computer to know how many pores are on your forehead?

Identifying a Retinal Scan

The *retina* is the layer of nerve cells at the back of the eyeball that acts as the projection screen for images passing through the cornea, iris, and lens. These nerves connect (via the optic nerve) to the brain, and convey the information that the brain interprets as vision.

Since the iris has to be open to allow light in for sight, and the cornea and lens pass light in both directions, it's possible to see the retina by just shining a light in there and watching it reflect. Optometrists and ophthalmologists use this as a non-invasive way to learn things about the health and function of your eyes.

As with many physiological biometrics, what were initially medical technology and procedures take on new uses as tools for biometric measurement.

Understanding the biometric basis for retina recognition

Although the primary functional purpose of the retina is to present the nerve cells to light so you can see things, all those nerve cells need support from blood cells, which are far easier to photograph than nerves. In 1935, Drs. Carleton Simon (no relation to Mike) and Isodore Goldstein discovered that the patterns of these blood vessels were unique to each individual and could be used to identify people. As with so many advances in science, this was discovered while they were studying something completely different (eye disease), but they were astute enough to see that this was an important find.

Unfortunately, digital imaging wasn't available in the 1930s, so identifying people using this method involved developing film and manually comparing the images to other images taken earlier. Not that surprisingly, the practical use of this idea in real-time biometric authentication had to wait until the mid-1980s.

With modern technology, we can take a digital image of the retina while the user's eye focuses on a specific point, and then use computers to compare the new image with a known sample.

Since the retina is well protected within the eyeball — unlike fingers or palms — it's not subject to wear and tear. The retina is also naturally stable; it doesn't change much over a person's lifetime. This uniqueness, combined with lifetime stability and protection from damage, makes the retina a very useful and accurate feature for biometric measurement. Figure 6-1 shows a cut away side view of an eye with retinal veins.

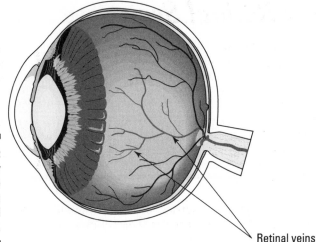

Retinal veins

Contemplating practical considerations

In practice, retinal biometric authentication and identification involves having you sit very still for several seconds while focusing on a specific point in space and a low-intensity infrared light is shined into your eyeball — where it images a small, specific capture area of the retina. Typically the light and camera apparatus must be no more than about 3 inches from your eye and shielded from ambient light. (Did we mention that you can't move your head at all during this process?)

Even though retinal scanning is considered very accurate, the use of this biometric seems to be waning as other accurate measures come into play — such as iris scanning, which can be done comfortably at a reasonable distance using ambient light. People who work at (or visit) highly secure facilities such as military installations or power plants can still be convinced to stick their heads in a hood and stare at the marker — but almost any other kind of facility is looking for quick, non-intrusive-but-accurate measurements.

Due to the way retina scans are acquired, some basic training is required to use these systems successfully; typically that training occurs when a person is enrolled in the system. This makes retinal scanning unsuitable for places that plan to authenticate or identify the general public — for example, ports of entry.

Where you will see retinal biometrics

Because of the limitations for distance, the training component, and the negative perception by many users of something that scans inside the eyeball, it's quite likely that you will not see retinal biometrics anywhere at all. Less intrusive methods that show characteristics not only better but from a distance — including better and more accurate eye-based methods — are available. Retinal imaging seems to be a dying technology.

There are still some very-high-security installations that have not converted their retinal biometrics gear to something newer yet — but we'd expect to see them adopt iris recognition just as soon as they can put the project and budget together to do so. It's the coming thing. As of early 2008, diligent searches for a manufacturer still making retinal imaging gear that *isn't* specifically for medical purposes turned up nothing.

Iris Scanning

The *iris* is the colored part of the eye that expands and contracts to allow more or less light in. For most people, this is the part of the eye that we remember most about someone, and is most commonly used as a general descriptor on such documents as driver's licenses and passports. In some ways, eye color alone has been used as a gross biometric indicator for centuries — "It was the green-eyed kid who took my candy. . . ." These days we have the technology to be more accurate.

Not long after Simon and Goldstein published their paper about the uniqueness of blood vessels in the retina, ophthalmologist Frank Burch proposed the idea that the iris was also sufficiently unique for identification purposes. The timing and development of usable technology for those purposes follows almost exactly the same path as retina identification: Aran Safir and Leonard Flom patented the concept in 1987, and John Daugman patented his algorithms in 1994. Thus a working iris-based identification system was born.

Due to the patents on iris recognition and the further patents on the algorithms for working with the iris images, owned by Iridian Technologies, all the available commercially available iris recognition technologies are created by or licensed by Iridian. As with any successful technology that solves a real world problem, there are a number of groups working on ways to accomplish the same task without infringing on the Iridian patents. No strong second player has emerged in this area yet.

Understanding the biometric basis for iris recognition

The basic concept behind iris biometrics is that the complex structures of the iris are unique to each individual and can be captured with a simple visible-light camera. A subject is positioned so the iris can be digitally imaged; after image capture, the Daugman algorithms are used to abstract the iris data and compare it to previously enrolled irises for authentication or identification.

If you've never done it before, grab a mirror and get a good look at the complexity of your own iris. You will see structures in the iris (see Figure 6-2) that range from radial to circular and even freckles. Something as seemingly simple as the base color of your iris is actually attributable to both pigmentation (in the form of melanin) and the texture of tissue and blood vessels in the iris. The structures you see are stable from about the age of one year; they don't change appreciably throughout your life, barring a disease that affects the eye physically.

Figure 6-2: Here's looking at you, kid: typical image of an iris.

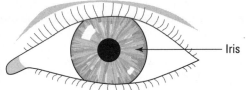

Iris

Contemplating practical considerations

Iris recognition may be the most promising of all the physiological biometrics currently in use due to its very high accuracy, non-intrusive sampling techniques, and reliance on only normally visible attributes.

Irises can be accurately imaged at distances of at least 3 meters, through glasses and car windshields while the subject is in motion. Iris-based identification has been done using old photographs taken several feet away and with the subject in a natural pose, but in those circumstances, the comparisons to known samples must be done the hard way: with humans doing the actual laborious comparison. Automatic comparisons are possible in more controlled conditions — but without the intrusiveness of stopping subjects and getting them to stare at a target while an infrared camera images their retinas from inches away.

Installed systems that have performed *trillions* of iris comparisons report perfect scores for eliminating false acceptance and false rejections. So far, iris recognition seems to combine the high accuracy of retina technologies with the ease of acquisition common to many behavioral biometrics.

In high-security, low-throughput applications, proof that the iris is attached to a living person is fairly simple to accomplish as well: Blinking on command and/or showing light-based pupil contraction and redilation would be hard to imitate convincingly.

Where you will see iris biometrics

Not too surprisingly, given the advantages of iris biometrics, many places are already using this technology — and many more installations are in the planning stages. The accuracy of the technology appeals to high-security applications such as military and national infrastructure; and its remote-acquisition capability and ease of use lend themselves to high-throughput technology and screening applications (such as airports and checkpoints).

The United Arab Emirates uses iris-recognition technology to screen all incoming visitors against a list of thousands of persons who have been expelled from the UAE. Border authorities have done 200 billion cross-comparisons between IrisCodes (the mathematical representation of the iris information) — and have caught 46,000 persons illegally attempting to reenter the UAE — with no false matches.

In the United States, the Child Project is an iris-based system for helping to identify and return missing children; as of September 2007, 1,400 sheriff's offices were participating. The company that supplies the technology for the Child Project (BI2 Technologies) also supplies Senior Safety Net and the Inmate Recognition and Identification System (IRIS).

The U.S. is using iris-recognition technologies in Iraq to control access to facilities, but has so far resisted the temptation to do more than capture facial images on passports.

Facial Imaging

As human beings, we learn to use facial images to identify people, starting with our first views of Mom and Dad. We get so amazingly good at this process that it's virtually impossible to fool us with a substitute face that's *almost* like the one we know — and we're fairly good at penetrating disguises such as false moustaches, beards, makeup, and the like, especially when dealing with a face we know well. As an example, Mike's 18-month-old niece can identify him from a photo taken before she was born — minus his new goatee.

We further extend our skills in facial recognition to interpret expressions with great subtlety and accuracy, so that we can say things like "That is my friend Eddie, and he looks happy." For the human brain, which has devoted lots of space and power to this task, the recognition task is simple; children master it long before they can speak, and it happens nearly instantaneously after they have it down. For some, the storage and recognition are so extensive that any face, once captured, will be familiar — even years later, when other contextual information about the person associated with the face is long gone. (Now, if we could just remember *where* we saw that person before . . .)

Computers, unfortunately, do not have the fantastically well-trained neurons that we do in this area, so they aren't nearly as good at faces as we are under most circumstances. Using specialized imaging hardware, computers are much better at seeing through smoke, costumes, and other disguises — but then they fall back on recognition algorithms that still fall far short of what your built-in ones can do.

Understanding the biometric basis for facial imaging biometrics

Facial-imaging biometrics are based, as you might think, on comparing information from two digital images of faces to see whether they match. The problem is that faces offer all sorts of challenges when you try to make direct comparisons, so you should note the careful language in the previous sentence — "*information from* two digital images" is not the same as "two digital images."

For a good example of why we can't just take two pictures and see whether they're identical, take a look at the angelic picture your sibling sent you of your four-year-old nephew. Now, hold that picture up and suggest to your nephew that he make the most horrible scary face he can, without using his hands or artificial fangs. (As a side project, take a picture of this new face and send it back to your sibling, titled "Reality.") Do you think that these two pictures would indicate these are even the same *species* when compared directly to each other? Typical changes to facial appearance that would throw off direct image comparison include (but are likely not limited to)

- ✔ Addition or removal of facial hair
- ✔ Body piercing
- ✔ Change in hair style
- ✔ Makeup
- ✔ Lighting changes from original image
- ✔ Different angle from original image

✔ Expression

✔ Tanning, black eyes, general pigmentation changes

Because it's really not possible to rely on directly comparing the images pixel by pixel, scientists have spent quite a lot of time figuring out how to represent the essential information in a face mathematically so that they can correct for cosmetic and imaging differences. The math is fairly intricate and not appropriate for this book, but we include a short description of the principles involved for the three main methods used in facial recognition.

Elastic Bunch Graph Matching (EBGM)

This method uses methods that closely match the fields in the visual cortex of higher vertebrates (including us). The mathematical model that we vertebrates seem to use is something called a two-dimensional Gabor function, which (you should be pleased to hear) we won't attempt to fully explain here. In effect, this method identifies local landmark features of a face such as the corners, top, bottom, and center of the eyes, assigns some characteristics of the image surrounding that landmark, and then compares these characteristics with a new set of landmark characteristics from a new image.

This method has some advantages over other current methods in recognizing facial images that are not oriented exactly the same way as the sample, but like the vertebrates it imitates, it requires some training to be able to work well with a given pool of faces.

Principal Components Analysis (PCA)

PCA uses linear algebraic techniques to reduce facial image information to the smallest set of uncorrelated components and then compares the distances between the features of two sets of facial image data reduced in the same way to see if they are the same. With respect to the reduction and "correlated components" consider the laugh lines near the corner of your mouth. If there is more than one on each side, the nearby ones are very similar in size and direction, so that whole feature can likely be compressed to a representative line that's more like an average of the two.

This technique allows the system to represent the necessary information for comparing two faces using very little information once the mathematical representations have been accomplished — which is nice if you have a lot of faces to store. On the other hand, it suffers a bit from the fact that facial images have to be *normalized* — meaning they all have to be the same size and the eyes, nose, and mouth in the sample images must be lined up before the PCA is applied.

Linear Discriminant Analysis (LDA)

LDA is a statistical technique that attempts to generate a good predictor of what characteristics a given face might exhibit, given a series of samples called a *class*. Essentially, the idea here is to toss out facial features that seem to vary greatly from sample to sample, and concentrate on those features that remain relatively the same. At the same time, LDA tries to choose predictors/features that maximize the differences between faces that are known *not* to be the same.

As with any statistical model, LDA's accuracy is heavily influenced by the class samples it's given to compare. If some of the facial image samples that are used to work out the predictor are not very representative of the face in question — that will have a negative effect on the predictor function and the system's ability to recognize faces.

Other methods

Given the complexity of computer comparison of facial characteristics, it might not surprise you to hear that there are actually many methods being explored for facial recognition, some related to the methods described in the preceding sections, some radically different. If you're mathematically inclined and interested in some of the other methods, look up Independent Component Analysis, Evolutionary Pursuit, Kernel Methods, Trace Transform, Active Appearance Model, and Hidden Markov Models. Nearly all these are even more complicated than what we've already described here. Maybe we'll include them in our next book, *Mathematical Algorithms for Computerized Facial Recognition For Dummies.* (We're kidding — we think.)

Contemplating practical considerations

Based on the methods required for recognizing faces, you should expect to use some very fast computers and possibly even some specialized processors to accomplish accurate facial recognition in a reasonable amount of time. Slower computers or faster response times will yield unacceptable false rejection or false acceptance rates (likely both).

Remember too that even the best facial-recognition system is not as accurate as something like iris recognition — or even fingerprint recognition. Nearly all current methods for recognizing faces have some way of accounting for poor capture angles, expression, and other issues of this nature. These problems are inherent to facial recognition; they make the kind of accuracy available in other methods impossible.

Where you will see facial recognition biometrics

Probably the best-known use of facial biometrics was at the 2001 Super bowl in Tampa Bay, Florida. The idea was that the Super Bowl was a high-profile target for terrorist attacks — and if officials were to capture images of 100,000 people as they walked through turnstiles and other checkpoints, then they could use facial biometrics to compare those images to a database of known criminals — and, in the event of a positive match, arrest them before they could start doing anything harmful.

Privacy advocates raised alarms at the idea of monitoring and identifying 100,000 private citizens without their knowledge or permission, but it turns out that both sides kind of had the wrong idea. Privacy folks didn't mention that to be recognized (identified) by any biometric system, you must first be *enrolled* — that is, your biometric information would already have to be stored in the system. 99.9 percent of the people entering the stadium were not enrolled anywhere, much less in a criminal database so their privacy was relatively unharmed. The officials needed to remember that under those capture conditions, you're lucky to *authenticate* a known face; actual *identification* is very difficult.

A live test of facial recognition at Palm Beach International Airport in 2002 failed to match volunteer employees (who had been enrolled in the system) about 53 percent of the time. Problems cited included eyeglasses, imaging angle, subject movement, and lighting. In a test that included 5,000 passengers and a database of 250 photographs, the system raised false alarms about two or three times an hour — and failed to identify *anyone* correctly.

Facial recognition is at its best in controlled conditions when comparing images taken under identical conditions. Although that sounds pretty restrictive, it's commonly used by law enforcement to compare mug shots to pictures acquired for this purpose — and, in some cases, to compare ID photographs from passports or driver's licenses to samples gathered in controlled conditions.

Upcoming Head-based Biometrics

There are a couple of head-based biometrics that are not yet mainstream, but have a fair amount of research backing them up — and interesting potential: ear and facial thermograph. Since you aren't likely to see them anywhere soon, we don't present any practical considerations (which would be purely artificial at this point). We do want to present the ideas here because this industry moves *very* fast, and we expect to see commercial systems using these methods eventually.

Recognizing ears

Some people look at ears, and some don't. We don't know how many times we've heard comments about the shape or characteristics of someone's ear and been completely in the dark, since we aren't ear people. Apparently the folks who *do* look at ears are on to something, though; studies show that ears are at least as rich in individuating characteristics as faces. Of course, since you don't smile or talk through your ears, they're easier features to get stable images of (unless, of course, they're covered by hair or a hat or adorned with little shiny objects).

The techniques and processes used for recognizing ears is nearly identical to those used for faces, so we won't delve into the biometric basis or practical considerations for ear biometrics as a section of its own. But note the advantages of ears: Their relatively stable appearance (earrings aside) and the amount of unique information they offer make them a better biometric than facial imaging, given the same circumstances. There is a great potential for using ear recognition as a multimodal biometric along with facial recognition to provide accuracy that neither method could produce alone. *Multimodal biometrics* use more than one biometric characteristic at the same time to provide greater accuracy for a biometric method that may be less accurate but in use for its ease of acquisition.

Ear recognition is fairly new on the scene and not well commercialized at this point, but the clear advantages over facial recognition in accuracy and ease of capture should move this technology along pretty quickly.

Recognizing facial thermographs

Another way of looking at faces (if you're a computer-imaging device) is to use infrared imaging techniques to get a picture of the thermal output from a face. This thermal output — essentially heat — is largely generated and controlled by the veins near the surface of a face; it can provide sufficiently detailed and unique information to serve as a biometric measure.

Unlike other vein-imaging biometrics, thermographic facial images are typically taken from a distance; the technology can't yet capture as much vein detail as (for example) hand-vein biometrics, which we cover in Chapter 4. At a distance, the thermal information from the veins is somewhat more diffuse, and this lack of detail can cause problems with accuracy.

Additionally, mood, health, exertion levels, and a number of other factors can change the thermographic map of a face — sometimes so much that collected biometric samples no longer match. Environmental factors (such as partial sun heating a portion of the face) can also throw off this kind of recognition.

With all these problems and capture difficulties, why would you ever use facial thermographs as a biometric? Well, the trick is not to use it exclusively. It's an excellent supplemental (or multimodal) technique when used with facial imaging; it doesn't rely on visible light, so it sees right through many disguises. A facial thermograph through a Halloween mask looks just like that same person without the Halloween mask.

Facial thermography also captures more than just identity. For some applications, the idea that you could know something about the mood, stress, or even alcohol consumption of the person being authenticated or identified is a plus.

Comparing Eye- and Face-Based Biometric Types

The following table provides a quick comparison of the eye and face biometric types that we discuss in this chapter.

	Strengths	*Weaknesses*	*Cost*	*Counter-measures*	*Convenience*
Retinal imaging	Accurate	Expensive, hard to train users, rare	High	Live guards to watch the process	Poor
Iris imaging	Most accurate of all current methods; easy acquisition	Only one vendor	Medium	Live guards	Excellent
Facial imaging	Similar to human process	Can be difficult to acquire; not as accurate as eye-based methods	Medium to low	Additional methods for greater accuracy	Excellent
Ear imaging	More accurate than facial, easier to acquire	Not mature yet	Medium to low	RFID ear tags	Good
Facial thermograph	Works through most disguises	Not accurate enough by itself. Must be used with other biometrics to increase accuracy	Medium	Additional methods	Excellent

Chapter 7

Other Types of Biometric Identification Schemes

*I*n addition to the biometric measures that we cover in the previous chapters of this part, a number of biometric measurements are not heavily used but show promise — for the future, in a specific area, or both. In a few special cases (such as brainwave and DNA biometrics), we discuss these measures strictly because if we left them out, you'd ask us why — not because they have any immediate practical value or promise for the near future.

When it comes right down to it, you can base a physiological biometric on the detailed imaging of just about *any* biological characteristic of a human being — and behavioral biometrics use almost anything we do often enough to become expert at, whether it's tying your shoe or snapping your fingers. For example, given enough sampling detail, we're pretty sure that the pattern of hair growth on your scalp would yield a sufficiently unique biometric to use for identification and authorization. Unfortunately most people would have to get their heads shaved before anyone could get a good look at the pattern. We're guessing a fair number of people would object to that (for the sake of a biometric measurement, anyway), so follicle-placement biometrics probably won't get a lot of attention any time soon.

Here we cover a lot of biometrics that may have potential problems just now — whether from a societal, technological, or legal perspective — but also have redeeming characteristics that make them interesting.

Recognizing Speech

Human speech is another one of those things that babies start learning how to tell apart early — starting with mom and dad and branching out to family and friends pretty quickly. When friends call us on the phone, we generally don't have to ask "Who is this?" We know very quickly from the sound of the voice who we're talking to.

For each of us as individuals, the list of recognizable voices is actually pretty long — including regular baristas, actors, musicians, radio personalities, and co-workers, in addition to friends and family. For most people who are reasonably active socially, the list expands to at least a couple of hundred people whom they can accurately recognize by just the sound of their voices.

Using this speaker-recognition capability to authenticate someone actually precedes widespread access to digital audio processing; earlier systems used analog filters to determine matches. In the mid-1970s, Texas Instruments built a prototype system that was tested by the U.S. Air Force and the MITRE Corporation. Once the cost of digital processing of analog voice signals became cost-effective in the late 1980s, many companies and academic study groups started working on ways to use speech for identification purposes. Take a look at Figure 7-1 for a graphic representation of a human voice.

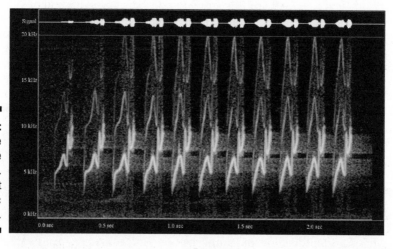

Figure 7-1: Voice sample graph (U.S. Government Public Domain).

Unlike most visually based recognition using analog images (which require computers to interpret relatively large amounts of two-dimensional image data before they can start comparing), computer voice recognition can use exactly the same sample as a human being would — or possibly more, through the use of sensitive microphones.

Understanding the biometric basis for speaker recognition

Speaker recognition is one of the few biometric measures that combine both physical physiological characteristics (your larynx) and behavioral characteristics (how you pronounce words). As it turns out, either one of these characteristics by themselves might be a reasonably unique biometric measure, so the two combined become a reasonably accurate biometric.

There are ways to make speaker recognition even more accurate by training the system using phrases that will then be used to match. For example, "My voice is my password, verify me" is a lot simpler to match if the system has learned how you say that exact phrase over a number of samples. Speaker recognition systems that try to know how you speak, no matter what you are saying, are more complicated and use way more computer time.

Physiological basis

When we speak, a surprisingly complex mechanism is activated that all work together to make the sounds involved in human speech. The physical size and structure of each of these structures contribute to the unique sound qualities of your voice. While it's tempting to think of the voice box or larynx and vocal cords as the only items that really influence how you sound, the human voice also uses the resonant spaces in the sinus and oral cavity to produce speech. It's pretty simple to test this — just plug your nose and see how the sound changes by just restricting the sinus. If you think back to the last time you battled the common cold, you'll recall the drastic changes in tonal quality that full sinuses cause.

The tongue and lips also physically affect how sounds are made, but more on the behavioral level than physiological. The ways your lips and tongue make sounds are learned behaviors — which you apply to match the ways you specifically grew up speaking (with possibly just a little influence from the shape and physical structure of your lips and tongue).

Speaker recognition is not speech recognition

They might sound almost the same, but while speaker recognition is focused on biometrically identifying someone by matching voice characteristics, *speech recognition* is the science of translating the human voice into text or commands. For example, you say "Mrs. Wright writes to right wrongs" and the computer prints "Mrs Right rights to write wrongs." *Speaker recognition* is the science of training the computer to recognize when someone says "Mrs. Wright writes to right wrongs" and it determines that's Mike Simon! Note also that speaker recognition is also sometimes referred to as voice recognition.

Behavioral basis

An old friend once told Mike that he spoke what little German he knows as if he'd been taught by someone from Atlanta. This friend was born and raised in Berlin and had lived in the United States for decades before she heard Mike butchering her mother tongue; she was accusing Mike of speaking German with a pronounced Southern accent, and not Southern Germany either.

The point is that the sound of our speech is a learned thing, heavily influenced by our learning environment. Mike's Spanish language skills were nurtured by spending years working alongside Mexican migrant workers in the strawberry fields of Oregon as a child, and while his vocabulary is still at a 3-year-old's level, Mexicans have told him that he sounds Mexican until he messes up sentence structure.

Our speech and language is influenced by geography, education, and even our jobs, to a point where a good human analyst can tell a lot about where you grew up, currently live, and potentially, do for a living by listening to you for a while. We'll discuss the language aspects of this later in this chapter in the section on linguistic analysis.

Factoring in heredity and environment

Heredity has influence over the physical factors that influence speech, and environment influences the rest. So, what happens when heredity creates a vocal apparatus that's shaped a lot like your father's and you also happen to learn how to speak from him? From personal experience, I know that from the age of 16 until I left home nobody could tell the difference between my father and me on the phone. Both of us would have to stop people who thought they were talking to the other. This pattern included employers, friends, and (in Mike's case) girlfriends and teachers; all of whom knew our respective voices quite well. For a detailed look at the apparatus that produces the human voice, take a look at Figure 7-2.

Twins: The people biometrics loves to hate

Throughout this book, you will see references to how various biometrics deal with identical twins. If you read a lot of literature on biometric studies, you'll no doubt imagine that biometrics researchers believe a small army of identical twins must be poised to descend on their precious idea and destroy it out of some twinly spite. In fact, most researchers are just using identical twins as the worst-case scenario for most biometric tests. For twins, the *bio* part of biometrics is nearly identical (barring physical damage) — and that can throw a wrench in the *metrics* part of biometrics (which depends on unique features). We're not ruling out the army of twins; we just don't have any real proof.

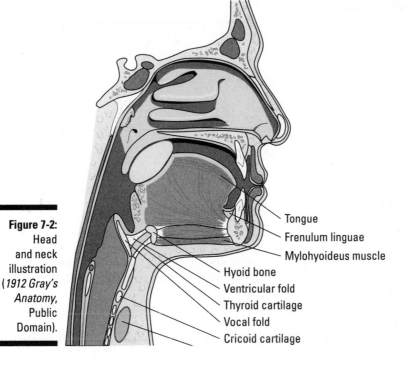

Figure 7-2:
Head
and neck
illustration
(*1912 Gray's
Anatomy*,
Public
Domain).

Tongue
Frenulum linguae
Mylohyoideus muscle
Hyoid bone
Ventricular fold
Thyroid cartilage
Vocal fold
Cricoid cartilage

We've never tested a voice-recognition system to see whether it would pick out the subtler differences between our voices, but the point here is that quite possibly the "identical twin" test would struggle a bit with speaker recognition.

Contemplating practical considerations

Speech has some interesting advantages that are hard to duplicate in terms of using existing hardware for good remote authentication or identification. For example, with a good phone connection it's possible to consider a system that verifies who is on the other end of a phone conversation using speaker recognition. If you're worried about someone capturing and replaying a passphrase, you could even ask semi-random questions and use general attributes of the voice rather than recognition of specific words.

Capturing high-quality speech samples is easy, using technology that has been undergoing constant improvement since the early wax-cylinder days. Any ten-dollar microphone is capable of recording sounds from 100 to 15,000 Hz, and the human voice ranges only from about 300 to 3,500 Hz. The harmonics of the human voice are highly individual, and based on the structure of the voice apparatus itself; they can range above 3,500 Hz, but not as high as 15,000.

Since capturing samples is pretty easy, and the analysis is usually limited to transformations of a two-dimensional waveform, the hardware required to do speaker recognition is inexpensive. Only these things keep speaker recognition from widespread adoption:

- ✔ It's easily confused by ambient noise, even when excellent noise-cancellation filters are used.
- ✔ It can be affected by transient conditions such as common colds.
- ✔ It has a tarnished reputation from systems installed over a decade ago that were not very accurate.

Where you will see speaker recognition biometrics

Speaker recognition is generally well accepted by people as a reasonable biometric to use; folks won't often object to its use (as they might with more intrusive mechanisms such as retinal or DNA sampling). Oddly, speaker recognition is more inherently subject to privacy abuses than most other forms of biometric identification; it can be used to identify people using their own spoken words. In Chapter 3, we take a look at the idea that free speech is impaired in an environment where you can always be identified by recordings of your very speech.

For a short while, something called VoicePrint was included with Mac OS 9, but reportedly it had accuracy problems; the false-rejection rate was around 40 to 50 percent in a room that wasn't absolutely silent. OS X doesn't seem to have anything like that included, possibly because of the reliability problems.

Even with some accuracy issues, the fact that speaker recognition can be accomplished via telephone is just too attractive for people not to use. To handle the accuracy problems, most folks who use speaker recognition at all are using it as one of multiple factors to authenticate a transaction. For example, you might initiate a transfer of funds online, and have the bank ask you for a phone number it can use to call you. The transaction then pauses until the bank's automated system calls the provided number — and uses speaker recognition to verify that the person answering is the account holder.

DNA as a Biometric Recognition Technique

Deoxyribonucleic acid (DNA), contained in each cell of all living things, is the basic blueprint for the organism. Starting in the late 1800s, scientists have been getting better and better glimpses of what DNA really is and how it's structured. In 1953, Watson and Crick published what is now known to be the first structurally accurate model for DNA, based on X-ray images taken by Rosalind Franklin (but sadly without crediting her).

Unlike the blueprints of a home, the DNA in our bodies is almost entirely unique to us as individuals, the exceptions being identical twins. For the 0.004 percent of the population that are identical twins, DNA will yield 100 percent false acceptance. For those same twins, biometrics based on phenotype — as determined by the interaction of genes with the environment in the uterus (for example, fingerprints or irises) — will differentiate correctly every time.

For the rest of us, our genetic material is unique and could be used to identify or authenticate us.

Understanding the biometric basis for DNA

In theory, DNA is a wonderful biometric measurement. It's known to be unique and the process of comparing one person's DNA to another is well known and not prone to errors like almost all the image-based biometrics. In theory, we could do a complete one-to-one comparison of one person's DNA to a known sample and have near-perfect accuracy. With certain exceptions (those darn identical twins again . . .).

To accomplish a DNA comparison, cells must be collected from the person being identified or authenticated, which are then processed to extract the DNA to be compared to a known sample. The cells to be extracted can come from blood, skin cells, or a swab from the inside of your cheek. Using current methods, the DNA is then extracted from the sample and duplicated many times so we have a large enough sample of DNA to examine.

Contemplating practical considerations

The current state of the art in DNA testing is really only useful for scientifically provable identification if hours or days can elapse while the identification is performed. Myriad — one of the companies that conducts genetic testing on the remains of 9/11 victims for identification purposes — said that the time to process a sample in its high-volume environment was about two weeks. A lab worker with nothing else to do but process a single sample could likely cut that time to hours, but nothing like a real-time turnaround is possible using current technology. So using DNA authentication to log in to your e-mail probably won't happen any time soon.

Another thing to understand about current methods for DNA comparison is that nowhere near the complete DNA of an individual is used to make a comparison. In fact, the most common method, *short tandem repeat (STR) analysis,* looks only for several series of repeating base pairs that don't even code anything genetically. These repeating pair sequences act as filler or garbage in the DNA; they're quite unique in length — repeating once in every 10 billion — when comparing all 13 recognized markers. Since this method doesn't even look at the part of our DNA that encodes genes, it's easy to imagine that a test using the rest of a person's DNA would be even more accurate.

Since the current tests take days or many hours at best, a more complete comparison that isn't just looking at the length of garbage or spacer sequences but actually looking at genetic encoding would almost certainly take months or even years with current technology.

DNA comparisons also require a sterile laboratory and trained technicians to provide good quality assurance for the process. Automated robotics are a part of any high-throughput DNA lab, but humans still are an important part of the process for now.

Where you'll see DNA as a biometric recognition technique

First, let's talk about where you will *not* see DNA used as a biometric measure. Until technology catches up with science fiction films — *Gattaca* (1997) in particular — you will not see DNA used for any authentication tasks. For real-world authentication, a wait of days, weeks, or even hours is typically not acceptable; it's highly unlikely that you'll see DNA used in this way.

Public or high-throughput identification is another area that DNA will likely not penetrate for a number of years due to the highly invasive nature of grabbing a few cells from you to test. Even if methods for making the comparisons were fast enough, it's unlikely that any large public group would agree to give up some cells, just to be identified.

Probabilities and DNA

The processes used in DNA matching are *deterministic* — given a specific DNA sample, they will always yield the same results when done properly. The outcomes, on the other hand, are *probabilistic*: The markers compared are not the complete DNA of either the known sample or the presented one, so duplicates that are not from the same person are possible (just not very likely).

You *will* continue to see DNA matching used to help identify persons — primarily from cells left at crime scenes — because the processes used to make the comparison are scientifically accurate and deterministic rather than probabilistic (as is the case with fingerprint matching). The accompanying sidebar explains why.

Gait-Recognition Biometrics

Have you ever seen someone in the distance and immediately known who it was, even without seeing that person's face? If you have, then there's a good chance that what you were subconsciously analyzing was the way that person was moving, and the movement itself was enough for you to determine who it was. Studies have shown that this talent is a fairly weakly expressed in human beings, but it does seem to be one area of basic biometrics in which computers outdo typical humans.

Gait is a biometric measure that is both behavioral and physiological. The way we walk is a learned trait, which makes it a good behavioral biometric. Gait is also somewhat dependent on physical characteristics such as the length of our legs, our weight, and foot size.

Understanding the biometric basis for gait recognition

As with most behaviors we learn to the level of "expert," the way we walk is distinctive to each of us an individual. For the first couple of years of life — before we develop the expertise we have as fully mobile, walking, and running people — it's very unlikely that any gait-based biometric would work. For some reason, however, we haven't been able to find gait biometric studies on toddlers anywhere.

Walking is hard

None of us really *think* about this after about 12 months of age, but the physical act of walking is tremendously difficult from a mechanical perspective. To move in a particular direction, you have to start leaning in that direction to the point where you would fall over if you didn't take a step. In essence, you are using dozens of muscles — along with all the bones in your feet, legs, and many in your back — to fall continuously in a specific direction while your feet move to stay underneath you. No wonder it's so hard to make bipedal robots that can walk reasonably well.

Once we learn how to walk and run without tripping too much, we settle into a rhythm and style of walking that is somewhat unique to us. This style accommodates physiological traits such as the length of your legs, ratio of legs to torso length, arm swing, and all the things that go into propelling you forward with all the grace you can muster.

Many of the things that are most telling — or at least most easily observed — about gait are best observed from a side view of the walker, where stride length, arm swing, torso position, and vertical bob can all be seen clearly. While the largest gross movements are most visible from the side, newer gait-sampling techniques are using more of the available information by looking at more than just stride and cadence.

Since so much of the body moves when walking, newer methods of gait recognition don't focus quite so much on the legs and stride; they pay more attention to how identifiable points on the body move in *relationship* to other parts of the body. This new kind of analysis tends to accent how closely gait is tied to physiological attributes, while still using many behavioral aspects as well.

Contemplating practical considerations

The larger movements associated with gait can be captured from great distances with reasonable accuracy, using no more than decent camera optics and lighting. Speaking of lighting, it's quite possible to capture usable gait-biometric information in total darkness using infrared cameras (though we're not sure that stumbling around in the dark would yield a valid biometric sample).

Newer gait-biometric techniques look at the movements of the body more holistically; they require more camera angles to capture everything they need for accurate comparisons. Many of the proposed installations that would employ this technique are for hallways or tunnels where people can be observed walking in a straight line for a specified distance.

Gait biometrics will still be on the new side for a while yet; accuracy in practical application has not been good enough for identification purposes when the pool of people is over a few thousand. Even so, many organizations and research groups are looking to use gait biometrics as a good screening process that might reduce the identification pool for other techniques (such as facial recognition or thermographic images).

Where you will see gait biometrics

Governments and intelligence agencies all seem fascinated with identifying people at a distance; no surprise that they are often the most interested in gait biometrics. Using (say) infrared cameras in low light to recognize gait from a small pool of captured biometrics is quite useful in covert surveillance. In anti-terrorist applications, being able to determine whether a target is among the group of people walking around a campsite from a half-mile away is another advantage of gait biometrics.

Outside of military uses, gait biometrics can be a good high-throughput screening tool — and so might be used in sports arenas or other large venues where a more intrusive biometric technique wouldn't be either feasible or well accepted. Even though gait in its current applications can't offer good identification from a large pool, it could inform crowd control to check on someone more closely if that person is a good match for someone in a smaller pool of suspects.

Closely related to stadium crowd control — but on a much smaller scale — is the concept of using gait biometrics to spot unwelcome returnees to a bar, dance club, school, or similar venue. Frequently people who have been expelled try to return and cause more problems. In these cases, there is no biometric collection of suspects, but a good video-surveillance system could provide the raw data and analysis to alert folks to a potential situation.

Typing Dynamics

This biometric measure has a long (and somewhat odd) history with one of the authors, starting in 1982 in university computer labs. While thumbing through some IBM manuals on the Display Management System for IBM 3278 terminals, Mike hit on the revolutionary idea that it would be possible to write a program to duplicate the operation of university login screens and capture logins from unwary students who would just see the screen flash, and then the login screen again. (Hackers do this today on phishing sites, but they probably didn't get the idea from Mike.) After avoiding expulsion, Mike gave some thought to how to keep someone from capturing a password in this way and using it to log in to his account. He came up with the wild idea that the *way* people type was probably unique to them, and if you could just watch the cadence and inter-character timing, you could see that the person typing in the password didn't just happen to know the password, but was the actual person associated with the account.

This marked the first and only time Mike ever hacked a system without permission — and the first of *many* times that his friends would later say, "Why didn't you patent that?" The story doesn't stop there, though. Many years later, working as a security consultant in Seattle, Mike met a very nice man named Gordon Ross who was at the time the Chief Technology Officer for NetNanny, the sole licensee of keystroke-dynamics technology (patented in 1989 by some very smart people at SRI). Gordon has been a friend and colleague since that day. The reason Mike didn't pursue that idea in 1982? It wouldn't have worked on IBM 3278 terminals, which sent whole strings to the computer at once as a block when an *attention* key was struck. Since the idea didn't solve the immediate problem, Mike moved on.

BioPassword, Inc., a spinoff company from NetNanny, still owns the patent for this technology; it has no commercial competition at this time.

Understanding the biometric basis for typing dynamics

Typing dynamics or *characteristics* are a form of behavior that can be used biometrically — but many of us learn them much later in life than most of the other behavioral biometrics. People in the *digital immigrant* class typically learned to type in high school (if they *ever* formally learned), while the *digital natives* who grew up with computers and the Internet learn to type as soon as their language skills support it. (They go on to learn to text-message and leave typing behind, but as yet we haven't seen texting dynamics used as a biometric tool.)

Typing may not be quite as deeply ingrained as walking or writing for many of us, but studies show that just two factors, the inter-character timing and the *dwell time* (how long a specific key stays pressed as you type) can yield 99-percent-accurate identification of the typist.

Contemplating practical considerations

Typing dynamics may well offer the most practical biometric available when computers are involved. That's because the sensor *is* the keyboard, combined with timing gathered using the CPU's own clock. No extra devices are required. To install this technology, essentially you just install the software.

The fact that this biometric technique does not require any kind of special hardware sensor beyond what's already attached to the computer makes it very attractive to online services such as banks and brokerages.

In years past, use of keystroke dynamics in an online environment would have required installation of additional software that would make it impractical for high-volume public use. These days, however, modern Web-browser environments such as Adobe Flash (98.8 percent of all Internet browsers according to Adobe), Java (84 percent), and JavaScript (nearly 100 percent) — can all be used to watch keystrokes while a user is typing passwords into a Web browser.

Factoring the odds

Multifactor authentication is a term you'll hear a lot when security folks are talking about secure authentication. The idea is that by combining more than one of the following, you can seriously decrease the odds that someone can acquire one of the factors (such as your password) and use it to get a false authentication:

✔ Something you know — a password or passphrase

✔ Something you have — an electronic token or key

✔ Something you are — biometrics measures

In large volumes, creating and disseminating the "something you have" part can be very expensive.

One possible method of biometric authentication — much easier to accomplish with keystroke dynamics than with most other biometrics — is what we'd call *continuous authentication and authorization monitoring*. The idea here is to continuously monitor the keystroke dynamics of someone logged in to a computer. You can check to see whether the user got up for a coffee and another person sat down at the keyboard and started using the system. You might be able to accomplish something similar with facial or iris recognition, or (for that matter) facial themography, but it would be considerably harder to accomplish, not as accurate, and relatively easily defeated. With typing dynamics, the usurping user would have to avoid typing anything to avoid detection. There are a lot of reasons an organization might want to keep track of who is typing on a specific keyboard at any particular time, including location or productivity tracking in addition to authentication and authorization.

Since keystroke dynamics measure the characteristics of *how* you type a specific sequence — say, a password — it's also possibly the only biometric available that's easily resettable. If a system focused on validating users through a password enhanced with keyboard dynamics, and someone learned to imitate the user's dynamics at login somehow, then simply changing the password would restart the learning curve for the hacker.

Where you will see typing dynamics

BioPassword has been making large inroads into online authentication with banks and brokerages, which are all under federal mandates to add more factors to password authentication.

Because it's a "sensorless" approach to biometrics, we can expect to see keystroke dynamics in places where good practice, regulation, or both may require something stronger than passwords alone. An example of this would be U.S. medical institutions that are required by HIPAA to record exactly who accesses a medical record. Keystroke dynamics can alert the organization to the fact that a password is being used by more than one person — or prevent access by anyone except the authorized user.

Interesting Biometrics Not Ready for Prime Time

In addition to the biometrics that we cover in the previous sections, there are a few "honorable mention" biometric types that aren't in current use in any except for very small specialty areas. We think they're interesting — even if these biometric measures may not be practical just now due to problems with sensor or sampling technology. They could easily get even more interesting when technology catches, and might even offer a perfect tool for solving a new or future problem you encounter. For now, they're just fun to think about.

Sentence-structure and linguistic-analysis biometrics

Statistical analysis of an individual's use of language is actually a pretty hot topic right now, but its uses are fairly focused on defense scenarios. The basic idea is to identify the person who wrote something by looking at his or her characteristic ways of using language — these, for example:

✔ How does the person put sentences together?

✔ What words does the person tend to use?

✔ How does the person structure ideas when writing?

These biometric methods are by no means exact ways of identifying individual writers, but they have interesting properties when applied to *groups* of writers and subject areas.

For really useful results, the analysis of sentence structure and language for this purpose needs large samples of known origin. For example, while it might be fascinating to see if Homer really wrote both *The Iliad* and *The Odyssey*, it's difficult to obtain a verifiable sample of his writing to use for comparison.

The modern age has yielded copious amounts of text in the form of e-mails that provide statistical fodder for language-based biometrics. Studies of e-mail collections have shown that it's possible to predict human relationships based on language clues in large bodies of e-mail (for example all the internal e-mail for a company) and even predict certain kinds of human behavior (like the tendency to betray one's employer and leak confidential data) based on e-mail language. Using these clues and examining the shared lexicon of groups, these measures can be used biometrically to identify *groups* of people and categorize the groups.

If all that sounds pretty *Orwellian*, you might want to reread the acceptable use policies for your company with respect to who owns company e-mail, who is allowed to read it, and for what reasons. In most cases, e-mail you send or receive in your company e-mail systems doesn't belong to you, and you're given very little right to privacy in it.

Brainwave biometrics

Brainwaves in the science-fiction *melt-a-stick-of-butter-with-your-mind* mode don't exist, so far as we can tell, but the human nervous system is electrochemical in nature, so pretty much anything we do — including thinking — does create a minute electrical field. These fields can be detected and even mapped using medical equipment such as electroencephalographs (EEGs), and magnetoencephalography (MEG) machines. Both of these measure electrical impulses generated by the brain directly. Here's what's currently in use:

- ✔ **EEG:** These devices measure activity by measuring electrical impulses passing though electrodes placed on the scalp, generally requiring direct contact with the skin in specific locations — requiring shaved spots in the hair.

- ✔ **MEG:** This is a relatively new technology that uses superconducting coils cooled to minus 269 degrees Celsius placed near the skull to detect electrical fields in the brain. These coils then induce a magnetic field in a superconducting quantum interference device, or SQUID, which can then be analyzed.

- ✔ **PET and fMRI:** Positron Emission Tomography (PET) actually measures the brain's consumption of radioactive glucose, and functional magnetic resonance imaging (fMRI) gear watches magnetically aligned hydrogen atoms in the brain react to a radio pulse. These devices measure brain activity without looking at the electrical fields. PET scanning can measure activity; fMRI can map out structures in the brain.

While all these methods yield unique biometric data, they're all either too expensive or too intrusive to gain much ground as popular biometrics (". . . to log in, think about *pumpkins* while we measure your brainwaves to confirm your identity . . .").

While technically not brainwaves, the electrical impulses your nervous system uses to keep your heart beating are measurable all over the body including the fingers. As it turns out, the harmonics of the electrocardial signals are a good candidate for biometric measurement; they can be measured as quickly and as unobtrusively as fingerprints. In fact, since it's possible to measure this electrical activity *from* the fingers, it makes a good proof-of-life for fingerprint readers so that the system knows the finger is attached to a living person.

Odor biometrics

We know, from observing bloodhounds, that it's possible to uniquely identify people by odor — and we know that it's almost impossible to mask that odor entirely without airtight seals. So odor has the potential to be a highly accurate biometric measure that cannot easily be masked or duplicated.

Unfortunately, this method has problems in both the technical and social arenas:

- ✔ Technically, reproducing the sensitivity and selectivity of a bloodhound nose and brain is very difficult. So far, we can't really come close with electronic noses, but work continues in this area.

- ✔ Socially, it's not particularly acceptable to tell people that you would like to identify them or authenticate them by using their body odor. (Imagine the enrollment process. Or don't. It probably wouldn't be pleasant for anyone involved.) Of course, this doesn't preclude the use of odor biometrics in a passive identification role.

Inevitably, odor is something that we leave behind everywhere we go — ask any bloodhound. Even the most pleasantly odor-free person leaves traces that trained dogs can identify and follow. Unfortunately, bloodhounds are not always convenient, and they do tend to shed on the back seat. There are agencies and police departments working with electronic noses to identify not only explosives, but traces left behind by humans. Some time in the near future we may see "smell evidence" presented as proof that someone was present at a crime scene. ("You know, something about this case stinks. . . .")

Comparing Biometric Identification Schemes

The following table provides a quick comparison of the biometric types presented in this chapter.

Note: The entries for sentence structure, brainwave, and odor are guesses because no established baselines currently exist.

	Strengths	Weaknesses	Cost	Counter-measures	Convenience
Speaker recognition	Easy acquisition	Easily spoofed unless counter-measures are used	Low	Use with speech recognition so that recordings don't work	Excellent
DNA	Deterministic, accurate	Takes a long time to process	High	Watch carefully for identical twins	Poor
Gait	Easy acquisition from a distance	Not very accurate	Medium	Use with complementary biometrics to increase accuracy	Excellent
Typing dynamics	No new sensors needed	Requires software on the devices	Low	Secure the transmission path to prevent replay attacks	Whatever is better than excellent
Sentence structure	Relates people to documents	Not that accurate	Unknown, maybe low	N/A	Excellent
Brainwave	Likely very unique	Politicians may be unreadable	High	Verify absence of brain transplants	Poor
Odor	Great names like "StinkID"	Manual verification unpleasant	High	Shower less, maintain smelliness	Excellent for user; poor for guard with "smell stick"

Part III
Implementing and Using Biometrics

The 5th Wave   By Rich Tennant

"Well, whoever stole my passwords was sure clever. Especially since none of my reminders are missing."

In this part . . .

This is where the rubber meets the road. If you intend to set up a biometrics solution in your organization, this section will help you establish selection criteria, run a pilot or two, choose a vendor, and build an implementation plan. After you get your biometric system running, you need to do some regular chores to keep it running smoothly.

Understanding and working with users is also essential to success when biometrics are introduced into an organization for the first time. There are misperceptions and fears that can be overcome with effective communication and training. You'll find it all in this part.

Chapter 8

Selecting a Biometrics Solution

· ·

· ·

*T*here are uses for biometrics beyond access control — such as law enforcement, surveillance, and other general identification purposes — but for this chapter, we focus only on selecting a biometric solution for access-control purposes. Frankly, all the other uses of biometrics are so specialized that if you're interested in them, you likely already have some background in that area.

Because we're focusing on access control, we should be clear about how we're using that term. In this context, *access control* is about using biometrics to manage who's allowed to use a particular resource. In many cases, that means access to computer systems, but it can also be applied to physical access — say, to an area of the facility via a door control, or even access to non-computer equipment such as a car.

Access control also has implications regarding how biometric comparisons are made. Because biometric comparisons can be used to either authenticate known users (as you would with a password) or identify users using nothing but the database with no other identification clues, we discuss how these techniques change the selection. As an example, consider a fingerprint biometric system that enrolls people using only the right index-fingerprint. Such a system is pretty good for authenticating users, given a cooperative user and a good look at the right index finger. If you happen to have the correct print, it's possible even to identify that user using his or her enrolled data as well. On the other hand, if you have all ten fingerprints, any of the prints can be used to identify that person. It may seem like a fine distinction, but it affects how you choose a biometric solution.

Identifying Selection Criteria

Before we start talking about the merits of various solutions, it would be a good idea to establish what criteria make sense to use in selecting a solution. There are quite a large number of biometric technologies and applications out there, so our first step has to be to establish what characteristics of the environment and the potential biometric solutions are of interest and useful in determining what to select.

The users

It's important to be familiar with the people who'll be using the biometric solution. At the very least, you must ask and have answers to questions like these:

- ✔ Are they members of the general public with a mix of ethnic and education levels and a variety of attitudes?
- ✔ Are they employees in a high-tech firm?
- ✔ Will they be using the biometric system several times a day, or just a couple of times each year?
- ✔ Will they be in a hurry and possibly be a little bit impatient (biometric controlled entry into a public restroom comes to mind just now)?

A great example of the need to understand users is in health care, especially urgent care environments. Health care is an area where convenient reliable access to information isn't just nice to have, it's an absolute requirement with lives hanging in the balance. It also happens to be a place where access control to extremely sensitive information is required by law and the expectation of patients, so biometric access controls are often considered for these environments.

In these cases, you have several groups of highly specialized user groups, each of which need access to essentially the same information for different purposes, sometimes in the same environment. An emergency room doctor might need to know about an incoming patient's allergies or prior conditions that they may access directly themselves, or indirectly through another emergency room worker. The doctor might be gloved, masked, and sterile in preparation for a procedure, while the nurse or other worker might be in the same condition, or sitting at a desk. Each of these users of the system have vastly different viewpoints and opinions about how the system should work while sharing the common goal of providing timely quality health care.

The environment

In this context, *environment* means the circumstances surrounding the contemplated biometric solution that will influence selection. For example, a high-throughput situation like airport check-in would be a bad place for something invasive and training dependent like retinal biometrics, but would be a fine place for iris biometrics since they can operate at a greater distance and with no user training.

Selecting the proper biometrics solution for your environment requires that you fully understand the unique requirements of the place where the proposed solution will be installed. Throughout the rest of this section, we discuss environmental factors that you should consider.

Accommodating physical requirements

There are a number of questions you should ask regarding the physical attributes of the place where the solution will be installed.

- Will this be happening indoors where lighting, temperature and other variables are controlled, or outdoors where things are more variable?

- What is the physical relationship (distance) between the authenticating user and the secured system or area?

- Is the authentication process attended by a trusted facilitator (guard or reception) or is it "self-service"?

- Are there attributes of the environment that rule out certain forms of biometrics, for example work gloves required and fingerprint ID?

It's helpful to try to describe all the potential users of the biometric system and describe how they'll use it to understand the physical requirements of each user and scenario.

Determining accuracy and "F"-rate requirements

Not all biometric measures are created equal in terms of accuracy and precision, especially given specific conditions in enrollment and subsequent use. Although iris recognition is extremely accurate and simple to enroll, it costs more and is harder to install than a simple fingerprint reader — and the fingerprint reader may be all you really need (even if it doesn't work well on cows). It's important when contemplating a solution to understand the False Rejection Rate (FRR), False Acceptance Rate (FAR), and Failure To Enroll (FTE) rates.

False rejection is annoying — it means a valid user has been incorrectly rejected by the biometric system, and must either try again or use an alternative authentication method. For most systems, you can lower the FRR simply by allowing the system more leeway in interpreting the specific characteristics of the biometric measures in use. The problem here is that by allowing the system to accept looser matches, you automatically increase the FAR — which means unauthorized users will incorrectly be accepted by the system. Adjust it too far and you might as well just prop the door open.

Commonly, medium-security installations adjust the system so the FRR and FAR are the same — an arrangement referred to as the *Equal Error Rate* or EER — but this isn't a cure-all. The appropriate FRR and FAR settings have to fit the security requirements of the installation. If higher security is required, adjust the system so the FAR gets very close to or at zero — and deal with people having to try more than once sometimes.

Because most systems allow you to adjust acceptance criteria, the key here is to look at FAR and FRR across the range allowed by the system and choose a biometric that matches the security needs required by the installation with as little false rejection as possible. Nobody likes being rejected, especially by a door or turnstile.

Unfortunately, biometric systems aren't equipped with big knobs labeled *FRR* and *FAR* that you can twist until you get it right. Instead, you'll be dealing with a greater number of adjustments — it's more like adjusting the amount of brown sugar in a cookie recipe. If you increase the brown sugar to make the cookie chewier and more resistant to drying out, you also need to adjust the white sugar to keep the sweetness right — as well as fine-tuning the baking time to deal with the hygroscopic nature of the brown sugar (metaphorically speaking). Bottom line: You have to tweak till it meets your security requirements.

Getting a better FAR might include modifying the software processing the incoming data so it rejects somewhat variant biometric data — but might also include more user training to get the best possible data from the sensors you have in place.

Cow network

A college project back in 1985 required that we identify individual cows electronically when they entered the milking stalls — which would then display information about each cow, retrieved from the database back in the office. For example, if the cow was on medication that would affect the milk, a notice was displayed so that particular milk would be shunted aside and discarded. The ID process itself was accomplished with proximity ear tags (which is *almost* biometric, since they become "part of the cow" in a sense.)

Because we also needed to check in the milking staff at each station (to track how they responded to the notices), we also provided them with proximity cards, shaped differently from those of the cows (and, oh yeah, not clipped to their ears). In fact, both cows and cowhands were using the same system, but had different physical interactions with it — and different needs based on their interaction. We wisely never told the cowhands that they were in the same database as the cows, just with different characteristics. We didn't tell the cows, either.

Understanding regulatory requirements

There are few federal or state regulations that address the use of biometrics specifically, but there's quite a lot of regulatory fussing about *authentication, authorization, and accounting* (*AAA* for short, which we talk about in more detail in Chapter 2) — so much, in fact, that we couldn't possibly make a definitive list of regulations (or even agencies). What we can do is provide some examples of industries that tend to regulate such things — and give you some of the specifics.

In general, the various regulatory bodies really care about AAA when users might need to be held personally accountable for actions performed while authenticated to a system. Industries or systems where that kind of scrutiny might be useful include

- Health care
- The Food and Drug Administration
- Law enforcement
- Banking and finance
- Any company managing data that includes credit cards

Although each of these industries cares a lot about who's using a system or has access to an area, the associated regulations differ wildly. For example, health care is regulated by the Health Insurance Portability and Accountability Act (HIPAA) Security Rule — which specifically *avoids* discussion of any specific technology in the area of "person or entity authentication." The FDA, on the other hand, requires (in 21 CFR part 11, an FDA regulation regarding the use of electronic signatures) that electronic signatures *not* based on biometrics must use at least two other factors to authenticate the user. Biometrics-based systems have no such requirement.

In many cases, regulations also require that the user not be able to repudiate their actions as an authenticated user. Essentially, the system must ensure that nobody using the system will be able to claim later on that "I wasn't there, nope, not me" when (ahem) certain actions occurred. In biometrics, this requirement relates directly to the false acceptance rate — and the possibility of replay attacks against the system, where a recording of valid authentication information is presented to the system. For example, in an installation that requires unattended authentication, voice-recognition biometrics would be especially susceptible to replay attacks — so they wouldn't meet some standards for non-repudiation.

There is no single answer for how to be compliant with regulatory requirements when you're using biometrics. In fact, only the newest regulations even acknowledge that biometric technology exists. Your best bet is to dig deeply into whatever regulatory environment you're working in — they're nearly always supported by online groups and professional organizations — and then apply what you've learned about the specific characteristics of each type of biometrics to your specific environment.

Pricing potential solutions

Pricing isn't *all* about product cost, and this is especially true for any product that affects large groups of people, like a new biometric authentication system. Clearly, normal information-technology standards apply — so you'll need to look at the expenses related to product, installation, deployment, training, and maintenance — all of which follow the same guidelines you'd use for any technology purchase and deployment.

Biometric installations add some specialized costs to the mix that are also worth considering:

✔ **Training/retraining:** If it's a new process that people will be using to access areas and systems that they're accustomed to using another way, everyone using the new system will require retraining — or at least allowed enough time to get used to the new system.

✔ **Enrollment:** Any biometric system requires users of the system to enroll by providing a known good sample of their biometrics to the system for later comparison. In most cases, that requirement calls for a human operator who can verify the user's identity and operate the system to enroll new users.

✔ **Productivity loss:** Although we like to think of enhanced security as an enabler for new business opportunities, introducing a new process with no direct addition to revenues *is* a direct cost to the organization. In some cases, you might be replacing an older, slower process — which means recording the resulting productivity gain is important.

Logging and reporting

Any system responsible for granting and denying access to important assets of an organization should be capable of keeping a record of access requests — and the system's responses. As with almost everything else in security systems, however, how much logging and reporting you do depends on the characteristics of the assets — and their protection profile (a technology and implementation independent description of the security requirements of the assets).

Typically, authentication systems record the following information:

✔ Who attempted authentication

✔ What that person was trying to gain access to

✔ When the attempt was made

✔ Whether the attempt was a success or a failure

Any good biometric authentication system records that data. With biometrics, some of these items show some unique aspects that we need to consider when selecting a solution. Here's a handy example: When a biometric system is working in *authentication* mode (as opposed to *identification*), the identity of the user is presented to the system, and then the biometric comparisons are made to authenticate the user. In the case of a failed attempt, biometric systems can record the biometric information *of the person attempting the authentication*. If it's someone other than the expected user, that information could later be used to *identify* a suspect, since you now have that person's biometric data in a special file, associated with the place and time of the attempt.

Considering the nature of biometric methods, *where* an authentication is attempted combined with *timing* can be an interesting and useful reporting tool as well. Because your biometrics are a part of who you are — and nobody we know has mastered the art of being in two places at once — a successful biometric authentication that happens in two widely separated places in too short a time always means (a) someone has compromised your authentication system or (b) your FAR is way too high.

Considering user privacy

A detailed discussion of biometrics and privacy appears in Chapter 3, but the issue is worth mentioning here as a selection criteria, too. Most biometric measurements are stored in an abstract form that allows the system to do a two-step process:

1. Perform a one-way hash that transforms the originally captured data to the abstract form.
2. Compare the abstract data to a known sample.

Because of those two steps, it's not possible to use (for example) fingerprint data in a biometrics database to create an *image* of the original fingerprint. The same is true for iris recognition, facial recognition, and most other forms of biometrics. If the users of the system know this, they're more likely to adopt the system — with a lot less hostility.

What can be done with the abstracted forms of biometric data is injecting that data into an authentication system at just the right place to successfully forge a biometric authentication. If a person's biometric data is compromised, that person risks some forms of identity theft *for the rest of his or her life*.

With that kind of long-term personal impact at stake, it's important to choose biometric systems that do a *really* good job of safeguarding the data they collect from enrolled users. In practical terms, that means encrypting the collected data — and using encrypted communications when transmitting it over networks.

Also worth considering, on the privacy front, is that many forms of biometrics can tell the operator more about a person than the operator really should know. Behavioral biometric information can indicate (for example) the onset of neurological disorders; sudden failures with most eye-based biometrics can indicate a long list of diseases — and medical information is generally presumed to be confidential, known only to patient and physician. Because disease or trauma-related biometric anomalies are nearly impossible to notice when they happen, plan on ways to reduce the impact these events have on the user and operator.

Considering standards and interoperability

Until recently, there were so few biometric products on the market that the very idea of interoperability wasn't interesting. Now, systems exist that store biometric information onto smartcards and passports that must be readable by a wide variety of systems, so both the American National Standards Institute (ANSI) in collaboration with International Committee for Information Technology Standards as well as the International Organization for Standardization (ISO) with the International Electrotechnical Commission (IEC) have published standards for the storage, use and transmission of biometric data. If you anticipate that interoperability among biometrics systems is (or will be) important to your installation, we suggest you take a look at the ISO/IEC Joint Technical Committee #37 (SC 37) proceedings and publications that can be found here:

```
http://isotc.iso.org/livelink/livelink?func=ll&objId=2299739&objAction=browse&s
        ort=name
```

Identifying the Field of Possible Solutions

After you've worked out the selection criteria most applicable to your environment, it's time to start narrowing down the list of potential solutions — by applying those criteria and zeroing in on your final list of candidate solutions. This is where you put on your healthy skepticism glasses and prepare to deal with sales critters. Don't get us wrong; some of our best friends are salespeople. Honest. And hey, salespeople with expense accounts are almost always good for a decent lunch — but to keep the playing field level, you have to let *all* of them buy lunch for you and your team.

Kidding aside, you'll most certainly need to talk to vendors *and* sales folks to get their take on what their products' capabilities are — and to hear what they have to say about their competitors. Just take both items with a few grains of salt.

The right way to start any technology project is to clearly express the requirements for the system you're contemplating. Generally, this process of gathering requirements involves these steps:

1. Speak to stakeholders about their expectations of such a system.

2. Compare stakeholder requirements to external requirements for the system, including legal regulations.

3. Write up the requirements for review by stakeholders, management, and representative users of the system.

4. Repeat Steps 1, 2, and 3 until all new input is analyzed and reviewed.

5. Incorporate feedback from stakeholder, management, and user review into the requirements document — and then translate the requirements document into a format for external consumption.

Speaking to stakeholders

The stakeholders for a particular proposed system or solution will be the people in the organization that have direct interest in the implementation, use, or outcomes of the solution. In practice, these are people in the organization who will benefit from the solution in some way, or whose work will be directly affected by the solution. For example, a typical biometric authentication system for accessing the computer server room would list the following people as stakeholders:

- ✔ IT management
- ✔ IT server staff
- ✔ Maintenance
- ✔ IT security staff

You have no real need to consult with sales staff or accounting in this case, since they won't be affected in any way.

Discussing requirements with stakeholders is always a bit tricky, since they don't understand the technology, and can't be expected to. Due to that lack of understanding, some of their stated requirements may be infeasible for any of a number of reasons — including technical capability, cost, or conflicts of interest with other stakeholders. The following list should help you collect the information you need without making any stakeholders feel they aren't being heard:

- ✔ **Keep notes about potential problems to discuss during the review process.** It's not a good idea to toss out anyone's requirements at this point, but you definitely should be making notes about potential problems and conflicts that you'll offer up in the review process. If you know for sure that 99.999 percent accurate iris scanning at a 12-foot distance while in motion is going to break the budget, make a note of that and bring it up later.

Please resist the urge to quash what appear to be infeasible require-ments in the information-gathering phase. At this stage, everything should be considered; telling folks that their needs won't be met really just discourages them from participating. You'll have plenty of opportu-nity to crush their dreams in the review phase.

✔ **Make sure stakeholders know the difference between wants and requirements.** If your stakeholders are prone to asking for everything they see on television or in the movies (especially science fiction), you might need to impress upon them the difference between wants and requirements.

A *want* is something that they believe would make the system work better or provide additional utility. A *requirement*, on the other hand, is something that, if not provided, will make the system of marginal or no use to them.

✔ **Collect stakeholder requirements separately.** When feasible, it's good to get stakeholder requirements from individuals or associated groups separately from other individuals or groups, so they aren't influenced by each other at this stage. Larger groups with differing interests can tend to drown out the quieter voices — whose requirements are just as real and important to the project.

✔ **Anticipate and address stakeholders' concerns about biometrics.** Because this is likely the first time you'll be speaking to some of these people about a biometrics project, you'll also need to keep in mind that the general public looks at biometrics with some degree of suspicion. It's collecting data about their bodies and behavior and doing who-knows-what with it. Try to anticipate their concerns and be ready to address them in these requirements meetings.

Be prepared to explain potential concerns to customers. If the project is about using biometrics for your *customers* (and not internal people), you'll need to work with your team to explain potential concerns to them, as well.

Analyzing requirements

The preceding section helps you compile a comprehensive list of require-ments from stakeholders. With that list in hand, it's time to make some sense of them — the requirements, not the stakeholders. Now is the time to docu-ment how the stated requirements each affect the solution choice, implemen-tation schedule, and costs.

It's also the right time to get a working understanding of any conflicts that the requirements may give rise to. For example, suppose you're preparing to provide your online customers with biometric authentication to enhance the security of your Web site. The operations-security folks are adamant that nothing short of iris recognition will meet their needs, but marketing and sales insist that whatever solution you choose must incur no additional expense or installation for your customers. If you check off all those customers who already have iris-recognition systems installed, that leaves you with approximately . . . yep, 100 percent of the customers needing new hardware and software to use the system. Uh-oh.

In such a scenario, you might be tempted to dismiss the iris-recognition requirement as silly or over-the-top — but resist that temptation. It wouldn't be surprising if, in the last phase, working with stakeholders who understand their requirements better than you do, you encounter somebody who states a need for something with the FAR and FRR rates of iris recognition. There are really only a few possible resolutions for this dilemma:

✔ One of the opposing requirements isn't really a requirement, but rather a want.

✔ One of the stakeholders isn't really a stakeholder, and that person's requirements can be safely ignored.

✔ You can heroically find a solution that makes the apparent conflict go away.

Under no circumstances should you decide on the first of these options without discussing it with — and getting support from — the affected parties. If they've said it was a *requirement*, they won't take kindly to seeing it dropped off the list without consultation.

Finally, here are a few more pointers to keep in mind while determining requirements:

✔ **Check external regulatory guidelines:** Another source for conflict may be external requirements imposed by regulations, laws, or guidelines. In this case, the winner of the conflict pretty much has to be the external regulatory body — unless your organization really feels like living on the edge and challenging that authority. The good news is that federal regulatory guidelines are typically so vague that opportunities for conflict are relatively rare — but it's good to check, just in case.

✔ **Avoid vendor salespeople who want to help you with your requirements.** Chances are their requirements will make their particular system the "only logical" choice. Better to develop your requirements internally without letting the vendors rig the game.

✔ **Do a basic sanity check on the resulting requirements list from the perspective of the person selecting and potentially implementing it.** The project requirements should be achievable with your current budget, in the available timeframe, and with acceptable interruptions to operations. From the perspective of an information technology stakeholder, these are among the requirements you bring to the table. You may have additional requirements regarding reliability of the system, maintenance, required training, and administrative usability that you should express at this time as well.

Reviewing your requirements

After you've come up with a good list of requirements for the system, you'll need to write them up in a format that others can understand. In Mike's case, this is the first time the requirements list makes it from his notes into electronic form — but your process may differ. The important thing here is to not get too fancy; just express the system requirements in a way that the stakeholders can review and comment on.

The fascinating thing about this part of the process is that it will almost always result in *additional requirements being requested* from various stakeholders. It doesn't matter how diligent you are in the first round of gathering requirements; there just seems to be something about the act of reading a "final" list of requirements that prompts people to remember the critical thing they forgot to mention last time. Knowing that, you can build in time for additional analysis and review of the inevitable new requirements in this phase.

Some people make this documentation task a carefully structured document with clear introductions, explanations, and conclusive explanations. Because this document is primarily for internal consumption, we usually just make it a list, starting with the words *This system must . . .* and saving the more detailed writing for the upcoming Request for Proposals (RFP).

Incorporating review feedback

After all stakeholders have had a chance to review the requirements, it's time to build the document upon which you'll base the RFP and subsequent implementation plans. Here's where you spend the time to give the project some context, carefully describe in detail the expected behaviors of the new system, and outline the impact to the organization's operations.

Often you'll notice that the feedback all has a particular tone to it — which you'll need to accommodate in the final requirements. For example, suppose that once everyone saw the complete list of requirements, each of them mentioned that people entering Building Seven — located in Barrow, Alaska — probably won't want to pull off their mittens in January and risk frostbite just to get a fingerprint scanned. If that's the case, and fingerprint recognition seems to be a good fit elsewhere, you may need to put the Barrow scanner in a vestibule to deal with that problem. (Barrow presents other problems with biometrics too, since at minus-70 degrees Fahrenheit people tend to cover ALL parts of their bodies, and even change their gait a bit to deal with the clothing.)

Remember back when we were describing "Speaking to stakeholders" earlier in this chapter — and we mentioned a stakeholder naming specific technologies (such as iris recognition) as a requirement? Most of the time, talking about specific technologies before you submit your requirements to vendors is a mistake. It's okay to talk about accuracies required, physical characteristics of the system, and users, but getting too specific about technologies or implementation too early is risky — especially when you're dealing with a relatively young and growing technology such as biometrics. You don't want to unnecessarily limit your options too quickly; you run the risk of eliminating great new products that would actually meet your requirements. Who knows — someone may have worked out the "running from angry polar bears" gait-recognition algorithms for the Barrow site, such that gait recognition meets your needs after all. . . .

Identifying and speaking with reference customers

If you have collected and analyzed requirements, and your analysis has resulted in a good understanding of exactly what you would like to accomplish, you have the selection criteria already prepared for vendors that you would like to submit an RFP to. It's the list of every vendor that could possibly meet your developed criteria. At this stage, if you don't know a particular vendor well and aren't sure whether its technology is a fit, send the RFP anyway — you might be surprised. At this stage, it's not your job to disqualify vendors; they will do that themselves as they respond to the RFP.

Once you have narrowed the field to a few vendor solutions, consider talking with people in other companies who have implemented those solutions. Ask the salesperson for reference contacts — preferably companies about the same size as yours, and not too far away in case you want to pay them a visit.

It's best to get at least two or three reference contacts from other customers, and develop a short questionnaire you can use so you ask all of them the same questions (after talking with the first reference customer, you'll probably discover several more good questions that you can then ask the other contacts).

If the salesperson can't or won't provide reference contacts, you might be just a little suspicious, unless your company is the first to use a new biometric product. In that case, get references from the same vendor for other, similar products. Remember that references aren't just about the technology, but how well the company does overall.

A very short list of questions that we use when talking with reference customers includes these:

- How well did the product work for you?
- Did you need assistance from the vendor? How responsive was the vendor?
- What problems did you encounter?
- What lessons did you learn?

✔ How would you do things differently if you could start over?

✔ Would you use the vendor's products again if you could?

Testing Potential Solutions

Testing biometric systems can be a tricky prospect; solutions can range into the hundreds of thousands of dollars, and many require integration into some sort of facility — say, the door controls or something equally hard to change just for a test.

With that in mind, there is still considerable testing you can do without tearing apart your facilities. Your goals in testing are to verify that the proposed system meets your requirements, and doesn't behave in any unexpected ways that you didn't anticipate. Grab a copy of your requirements, and we'll see what we can do.

Vendor/manufacturer on-site testing

Although it may not be feasible to install a six-camera gait biometric into your lobby just to test a potential solution, you can bet that the manufacturer will have facilities set up for you to visit to get a feel for things. Even better, they might have a cooperative, happy customer (you should be asking for happy customers anyway) that would consent to showing you their facility and talking about implementation, maintenance, and their general impressions. It's not the same as setting it up and letting your own users test, but in some ways this kind of test will offer up better data than you could in your own facility.

Getting mature data

A test setup is "only a test" — something that's been artificially created for the purposes of seeing how a proposed solution works. Unless you have a gigantic budget and a healthy implementation schedule, it isn't practical (even possible) to set up a system that will work exactly like the final product long enough to allow people to get trained, use the test system long enough to get past training time, and replace their general unfamiliarity with a routine.

By looking at an already-installed system, interviewing users and administrators of the system, and watching it in operation, you should get a good feel for how your installation would work after the newness wears off and things settle in a bit. A really great manufacturer or vendor will have at least one or two satisfied customers who can spend a little time with you — and a longer list of other customers who are willing to give the product a thumbs-up over the phone.

At this stage of testing, we also ask manufacturers for customers who are *less* than happy with the product, though we don't always get useful replies to that one. The least useful reply is

> *Everyone loves our product, we don't know of anyone who would say otherwise.*

A slightly better reply is

> *Well, XYZ Co. didn't care much for the "running from angry polar bears" gait-recognition system, but it's because all they had chasing them were slightly upset geese.*

And the best response of all is

> *We do have some folks who haven't been entirely satisfied with our product who may be willing to share their experiences with you. We'd like to have the opportunity to explain what happened with them after you've made contact.*

Any manufacturer confident enough in the quality of its product to put you in contact with less-than-completely-satisfied customers is either really good at bluffing, or pretty sure it has a great product. In any event, you should be able to get a good enough read from the less-than-happy customers to classify them as

✔ Curmudgeonly and dissatisfied with almost everything, including your phone call

✔ Unreasonably focused on some aspect of the product that isn't important, or at least isn't important to you

✔ Legitimately concerned with some aspect of their interaction with the company or the product

Using industry data

In place of actual hands-on testing, testing by industry groups can be useful for establishing quantitative measurements of products to see whether they meet your requirements. Groups that offer such testing include the International Biometric Group (IBG) and the National Biometric Security Project (NBSP). Their tests primarily determine accuracy, FRR, FAR, and other easily quantifiable measurements, but aren't quite so useful for evaluating user experience, impact, or other less quantitative measurements. Note that performing this kind of testing is expensive; in many cases, the testing costs are borne by the biometric device manufacturer, the person needing the results (you), or sometimes both. That shortens the list of tested devices, but if one of your prospects is on the list, it's helpful data.

The NBSP specifically makes Performance and Standards Conformance test results available on their Web site at www.nationalbiometric.org. The specific performance measurements available are False Accept Rate, False Reject Rate, Failure To Acquire Rate, Failure To Enroll Rate, and Throughput Rate. Because each of these measurements is a likely candidate on the technical side of your requirements document, this testing can be quite useful.

NBSP standards-conformance testing includes both the ISO and INCITS standards mentioned earlier in this chapter (in the "Considering standards and interoperability" section), as well as some industry specific standards from the International Labour Organization (ILO) and the International Civil Aviation Organization (ICAO). If compliance with standards is on your list of requirements, take a look at their testing in this area too.

Keep in mind too that each manufacturer will most certainly have test data as well — usually describing how amazing their product is, and directly comparing it to the unreasonably inferior competitive products. Oddly enough, each of the competitors will have similar test data that paints the same picture for their products. Don't be tempted to just ignore what each manufacturer says about its product — and pay *close* attention to what they say about their competitors. Most manufacturers won't lie outright; there is usually some kernel of truth in the claims they make for their stuff, as well as their take on what the competition is up to. If nothing else, it provides conversation points when you have your next vendor meeting.

Testing installations

Sometimes it's actually feasible to install a system for testing purposes, either because it's something simple (such as a fingerprint reader for authenticating to a computer system) or the project is big enough that spending money on a pilot is worthwhile. In these cases, getting value out of your testing requires a fair amount of planning and some attention to detail in collecting and interpreting results.

Defining success

First, you need to decide what outcomes from the testing will be needed to deem it a success. Clearly, this would include meeting all the specific requirements you've already developed, but are there additional criteria (in the "want" category) that you'd like to evaluate in the test environment? Are there factors that seem obvious, that didn't make it into the requirements, but will clearly be a part of testing — such as, "the system operates continuously with at least 99.999-percent availability for the duration of the test?"

Careful attention to very specific, measurable success criteria is important because (a) it makes the decision/selection process easier in the end and (b) sometimes the manufacturer will partially fund or support the pilot project in some way — say, by loaning equipment and/or offering free support. In such cases, the manufacturer will want to understand very specifically what criteria must be met for their product to be selected. Often the sales critters involved will get pretty pushy about this, calling daily to see whether the acceptance criteria have been met and you're ready to sign a check. When this behavior gets annoying, we usually just explain that the CEO is concerned about the amount of time we've spent on the phone to support this product and wants us to explore other alternatives unless we can lower the contact time to zero for a month or so. If that doesn't work, just tell the salesperson not to call you for a while.

Also note that when you're gathering data in areas that the manufacturer already publishes — such as FAR and FRR — significant deviations mean that either the manufacturer was very optimistic in its testing, or something is wrong with your test installation. Both are good reasons to have a representative come in for a closer look at your setup.

Selecting test subjects

Getting the right mix of people into the test program is also vital to getting good results from a test installation; you want to get usability information from a wide sample of the intended user base. All too often, technology tests are conducted using other technologists — whose view of how systems work is completely different from that of normal people. The right mix will include people who have varying degrees of familiarity with technology, biometrics, and authentication.

A somewhat delicate subject to address while choosing people for biometric testing is that you'll need to choose people at extreme ends of the physical spectrum related to the specific biometric to know how the system performs with the extremes of the physiological characteristics it's measuring. For example, Mike would bring in his 7'1" co-worker to help test the palm-print scanner system, since this guy's palm is about 1/3 wider than Mike's and the new system should be comfortable for everyone to use. (Now, if *we* were designing a biometric access system for NBA locker rooms, we'd find the 5'2" custodian and make sure the vertical placement of the sensor still worked for him.)

Be sensitive in this part of the selection process; don't include users who will obviously not be able to use the system, unless you're also testing the alternative authentication system. If your palm scanner is right-hand-only and you have someone in the company with no right hand, you already know that person can't use the new system, and you'll have to make alternative accommodations.

Training staff and users

Because you're going to have real people using a real system, you'll need to actually *teach* the selected users how the new biometric system works, and give them a taste of how they'll interact with it. For the user experience to be well supported, you'll also need to train administrators and support staff so they know how to operate and maintain the new system. You might be tempted, in some cases, to skimp on training administrative and support staff for a test — but your results will be somewhat skewed if you do, because the users won't be getting the kind of support from your staff that they ordinarily would.

A good example of how staff training helps is in the enrollment process. For users to start using the system, they'll have to go through an enrollment process that captures their biometric data and stores it in the system for later comparison. In many cases, enrollment requires the acquisition of multiple samples to have enough data to eliminate mismatches that come from poor positioning or other factors in normal use. As you might expect, running your fingers though the print reader ten times in a row is a little tedious, and poor training of the person facilitating enrollment could make it unbearable or even invalidate the enrollment process altogether and force users to go through it all over again. Nobody's going to like that much.

While training administrative staff and normal users on the system, pay close attention to how well your training material is received. Are any emergent problems in training due to poor materials, or is the system actually hard to learn? Training will be an ongoing part of using and maintaining the solution you select, so understanding how well this part of the process goes is a critical part of evaluating the proposed solution.

Flipping the switch

Once the solution is in place, your guinea pigs — (ahem) *users* — are selected and everyone is trained in the proper use of the test system, it's time to hit the big switch and actually start using the system as a part of operations.

In some ways, this is harder to do with a test installation than it would be if you converted the whole organization. That's because a test installation uses two sets of rules and operational procedures — one for the test subjects and another for everyone else. Keeping track of who should be doing what (and when) can be somewhat time-consuming, but it's important if you want useful test data.

Make sure everyone involved knows what to expect when the test system is operational, and to whom they should report unexpected behavior. Also be sure to allocate time to monitoring the operation of your new biometric system — long enough to feel comfortable with its level of operation and stability. The first day, or even week, of operation for your test installation is not a good time to catch up on your vacation days. Instead, make sure that all the places where you want to capture information about the operation of the test are actually reliably capturing data — and that the data makes sense.

Measuring results

Once a reasonable amount of time has elapsed, you'll have the data you need for making a final decision regarding the system under test. What amounts to a "reasonable amount of time" depends on how long it takes to make sure that each of your requirements for the system has been tested — and that you have data supporting the success or failure of each one. Even if you've covered all your requirements in a relatively short time, it's sometimes helpful to run a test system through one or more natural cycles specific to the organization. For example, if you've implemented a palm reader that limits access to enrolled users only during their actual work shifts, you should operate the system through one complete combination of shifts, as well as any potential reorganization that might move people from one shift to another.

The final analysis will have one fairly simple component — essentially a checklist with all your requirements on it and a column for yes, no, and comments. The more difficult part of measurement is really all the other information that you acquired while operating the test, some of which you weren't even looking for. If you replaced a human guard at the door with a biometric-system-and-a-man-trap (a system which prevents a person who fails to authenticate from running away by *trapping* them), and everyone complained that the lobby just seems too barren without a real human greeting them in the morning, how do you figure *that* into your analysis?

Making the Selection

You've vetted your requirements and you're pretty sure any solution that meets them will be acceptable. You've documented these requirements and received proposals from appropriate vendors, describing products that not only meet your requirements, but calculate the 371-millionth digit of pi (it's 3) while not busy improving your bottom line by at least 15 percent. What the heck, assume you've had the needed budget, time, and expertise available to test one or more of these potential solutions, and have copious data available to make a final decision on your new biometric authentication system.

In the best of all cases, you have identified several potential solutions that all satisfy your requirements — including functional requirements and budget. If you didn't include the stability and support potential of the manufacturer in your requirements or RFP, you should definitely look at that now as a final deciding factor.

It's no good to install a whiz-bang biometric ear-recognition-at-a-distance system that meets your needs perfectly if the company goes bankrupt next month or has a poorly organized support staff. That kind of critical information can be difficult (or impossible) to come by, but in the case of large installations involving serious cash outlay, it's quite typical to request financial statements from vendors or manufacturers to determine long-term viability.

If you have several viable candidates, the world is your oyster and you can start haggling on price. In fact, in a competitive market where you have several viable choices, the final selection frontier *is* price. We're not saying price should be weighted more heavily than other criteria; that depends on your specific circumstances. Rather, within your original budget constraints, pricing is the last item you use to help make your final selection. That's because it's a variable that depends on the final list of candidates and can be influenced at the final stages of negotiation.

What if your process yields only one viable candidate solution? At first glance, it may look like the decision is already made and you should just proceed. On the other hand, maybe your requirements were overly restrictive. There are a couple of factors that you should really consider if your process only yields a single viable candidate. If there's only one player on the field, that vendor's pricing is based on lack of competition; it's hard to effectively negotiate at that point. Also, if that candidate is the only one to meet your criteria, guess who pretty much has you over a barrel in years to come, when it's time to consider pricing, providing maintenance, and delivering upgrades? Still, if you're still comfortable with the sole survivor at that point, your choice is made for you.

Playing nice

In a competitive-bid situation, you should always consider explaining to the losers why they didn't win. If you're respectful and considerate in how you present this information and the bidder is smart, they walk away with valuable information about how to improve their product or their response to RFPs. Occasionally, you also learn things about their products and the winning bid that you would never have known without the follow-up. ("Yeah, everyone seems to be upset by the burning sensation as we laser-image the fingerprint. We're working on a lower-power laser for the next version.")

Chapter 9

Implementing, Supporting, and Maintaining a Biometrics Solution

Implementing a biometric solution is like a lot of big IT projects, but there are a lot of wrinkles that most IT people aren't used to dealing with. Although a biometric project involves fundamental changes to authentication and authorization services on the network, implementing biometrics also requires changes in users' behavior that you don't see in most projects — and collects fundamentally personal data from users that they've never had to share with the company (or almost anyone else) before.

Likewise, maintaining and upgrading a biometric system comes with unique (and sometimes unexpected) challenges. Human factors come into play here more than in almost any other type of IT project, because of the unique ways the users interact with biometrics. People tend to be particular about what they stick their fingers or eyeballs into; you have to recognize that while considering your implementation. Success or failure in most biometric projects is more people-dependent than technology-dependent.

Implementing the Biometrics System

At this stage of the process, the biometric system has been selected and perhaps even purchased. Here we depart the realm of the theoretical and enter the world of reality — of putting your money where your mouth may have been during the selection phase (even if it alters your basic facial-recognition biometric sample). It's time to make the chosen biometric system *work* in your organization.

If you're the person who is supposed to make it all happen, perhaps you're feeling a bit like a mountain climber standing on the flanks of Mt. Everest, gazing at the wind-wracked summit over three miles higher than your relatively comfortable vantage point and thinking to yourself, *How in the world am I going to get from where I am to that foreboding place up there?*

Thankfully, (believe it or not) most biometric projects are far less effort than climbing Everest. It all starts with a plan, and help from others in the form of expertise and support. And think about how nice the view will be when you have the entire system up and running.

If you're a small company that's just getting a new lockset for the computer room that includes a biometric fingerprint reader, you can probably just buy the lockset and make sure that all the people who need to get into the computer room get registered so they can get in when they need to. No big deal. You probably don't need a lot of planning for that, beyond making sure that the FAR and FRR are within the tolerances you need. But if you're implementing a biometric building-access system in a twenty-thousand-employee organization, you'll need to do a *lot* of advance planning to make sure the project goes off without a hitch.

Building a plan

A biometrics project — especially a big one that affects large numbers of users — requires advance planning to ensure that the implementation team will be able to design, build, and activate the biometric system. Planning is also required so the implementation can be completed on time and on budget, and the biometric system will perform as originally expected.

Getting a good project manager

Hopefully an experienced project manager is available to build the implementation plan. The larger the project, the more vital it is that a project manager with experience in project planning and management be available to keep the implementation moving in the right direction.

We're reminded of a project manager who ran a large, year-long, single-sign-on project team. Jeff ran a tight ship: he made sure we showed up at our weekly project meetings, that we were making progress on our tasks, and that he was aware of all significant issues. Left to our own devices, we'd have lagged behind and become distracted with the other cool and interesting technologies that were vying for our attention. Jeff kept the schedule and budget updated, helped overcome issues, and kept us focused on the project schedule and our need to pay attention to the project.

Sadly, not everyone *appreciates* great project management. There are always a few people in any large project who feel that the communications requirements are excessive — that the schedule is more like a "guideline" (they wish) and their own part of the project is essentially independent of the rest (ditto). You'll hear them grumbling in the back of the weekly meetings about having "real work" to do, and asking if they really need to meet the deadline for their part as long as it's done before the final installation date. Unfortunately, there's no cure for this mindset except time and experience. When possible, it's fun to have them project-manage a small group of their own on projects that can afford to be a little late (just be sure to tell them that the schedule is tied to their bonus).

Putting the plan together

A savvy project manager won't build the schedule in a vacuum; instead, he or she will ask members of the project team (and other interested parties) plenty of questions to understand what's needed to get the biometrics system implemented. Here's a short list of what the project manager has to do:

- ✔ **Identify tasks:** The project manager needs to know: What are all the tasks required to implement the biometric system? This quest should start with the high-level tasks, but should include the details as well.

- ✔ **Identify hardware resources:** This should include any servers required to support the solution, as well as any additional hardware needed for existing systems. For a biometrics installation, this will almost certainly include specialized hardware that isn't on your "currently approved" list.

- ✔ **Identify software resources:** Any additional software required to implement and support the biometric system should be identified. This should include new software as well as upgrades to existing software.

- ✔ **Identify network resources:** Chances are your biometric solution requires some sort of network communication, particularly if it is used to support authentication in an IT environment. Not only do you need to know what network resources are required, but also how much network traffic your biometric system is expected to generate.

  In most cases, the biometric identification process needs to happen almost in real time — which may have an impact on network design and the proximity of biometric databases.

- ✔ **Identify staff:** Projects require trained and experienced staff members who know how to perform the various tasks that have been identified. If internal staff aren't available — or aren't experienced in one or more areas — you may have to identify contractors or consultants who can fill the gaps. In an ideal world, contractors manage themselves, but keep in mind that you'll have expenses related to managing them as well, whether in their up-front fees or your own staff time.

 ✔ **Build a budget:** After you've identified all the needed staff, hardware, software, and other resources, you can establish a budget. It should address all the expenses shown here, plus those not covered in this section.

 ✔ **Build a schedule:** When most of the preceding items are figured out, the project manager can put a schedule together. The schedule should include all the tasks required — including interdependencies, prerequisites, and the resources required for each.

The plan in its entirety (and by this we mean the required resources, personnel, schedule, and budget) must be subjected to executive review and approval. Only after the plan has executive blessing can you safely proceed and execute the plan.

Executing the plan

Successful projects are not judged by their excellent plans, but by excellent execution. Many teams can create good plans, but fewer can carry them out. The inclusion of biometrics — still a new technology that incites factors that many IT departments are struggling with — can be enough to jeopardize the success of even the best plans.

Okay, we're not telling you that your biometric project is going to fail — far from it. If you're reading this book in a bookstore (or reading an online excerpt), know that we're not telling you to shy away from biometrics because it's too challenging; in fact, we are enthusiastic about biometrics and the increased security that results from their use. Rather, we're saying you have to know where the risks are — and anticipate them.

Good planning is as much about anticipating potential failures and avoiding them as it is about finding the path to success.

Biometric-related risks

The specific risks associated with a biometric-related IT project are entirely due to the unique aspects of biometrics. Here's a rogue's gallery of risks:

 ✔ **Changing human behavior:** The introduction of biometrics into a technology environment demands that people change their behavior in a fairly radical way. With new applications or tools, we're asking users to click *here* instead of *there*, to select data *this* way instead of *that* way, and so on. But a biometric project demands a *fundamental* change in behavior. Imagine replacing all the light switches in your home with overhead pull-chains. How long do you think it would be before you stopped reaching for the wall switch every time you walked into a dark room? Okay. Multiply that hassle factor by a factor of at least ten.

✔ **New personal behavior:** Using biometrics requires that users interact with technology with their *bodies*. But wait — keyboards and mice are handled with hands, and displays are viewed with eyes. But people are accustomed to that. Biometrics introduces a new behavior that many users have never done — or have done rarely. Now we're asking users to subject a part of their bodies (fingertips, palms, irises, retinas, whatever) to the system. Many will feel that use of the new system is . . .

✔ **A "personally invasive" issue:** Biometrics measure what some users consider personal — and private. Users may not wish to share their fingerprints, iris scans, or other close measurements with the organization, feeling that it's an invasion of their privacy. Some will openly object; others will quietly object but submit begrudgingly.

✔ **Gunk in the electromechanical equipment:** IT departments are becoming accustomed to (nearly) trouble-free electronic equipment such as sealed disk drives and optical storage. Many of us started our careers back when computers had a significant mechanical aspect to them: card readers, line printers, reel-to-reel tape drives, and paper-tape readers required a lot of care and daily cleaning. Although biometric devices may lack the gritty mechanics of a card-punch, they *are* physically handled by users — and more prone to just wear out and/or get dirty than most IT departments are used to. In practical terms, we're saying that many types of biometrics technology require some sort of periodic cleaning and/or maintenance. The bottom line is: Who is going to add "janitor" to their job description?

✔ **Fear of communicable disease:** Biometric devices that involve physical touch (such as a fingerprint reader or palm scanner) may be a put-off for some users. With the recent public concern over bird flu, TB, and MSRA, more people are wary of putting their hands (or other body parts) in places where others may have left germs. Hand-sanitizer stations nearby may help — but this is just another one of those factors that can make or break the implementation in some cases.

Project management

Project management doesn't end when the project plan is built, finalized, and approved. That's when it *begins*! Gas 'er up and let's go: the real skills of project management are put to the test *when the project begins* and the project manager has many issues to deal with, including these:

✔ Project team members whose other projects are still going past their anticipated end, which makes them less available for this project.

✔ Project team members who are distracted by less important things.

✔ Project team members who are distracted by more important things.

✔ Unscheduled sick time. (As opposed to *scheduled* sick time? We won't go there.)

✔ Vacations — especially near year-end in organizations that enact use-it-or-lose-it policies.

✔ Low estimates on effort required to complete tasks.

✔ Transfers and resignations.

We think you get the idea. A good project manager can't squeeze blood out of a turnip (although we have worked with some who've tried), but one can do an admirable job of keeping people focused and working as close to the original schedule as possible. Later on you can try to find a market for turnip juice.

Really great project managers will do a good job of showing everyone how important it is that their part of the project goes exactly as planned — and how critical their efforts are to the project and the organization. A team member who understands this and still doesn't perform should really be doing something else (maybe some*where* else, if you catch our drift).

Mid-course corrections

Surprises happen, even in the most disciplined organizations. Sick time, vacations, vendors changing the specs, servers arriving late, resignations, and other project emergencies throw a monkey wrench in projects, making them take longer and cost more (or, getting done on time while cutting corners that you'll pay for later after the project team and the subject matter experts disband). But these things happen to almost every significant project — a good project manager expects them and deals with them, while keeping the project as close to the original schedule and budget as possible.

Running pilots and tests

There are a few ways to approach a technology implementation project. When an organization is bringing in an entirely new technology, we suggest that the organization try it out before making the purchase. It's a little easier to change your mind when money hasn't yet changed hands. You can try out a vendor's biometric solution in some limited capacity — and if you don't like it, you can put it back in the box and tell the vendor, "No, thanks."

Unlike the white lies many of us tell when we return unwanted merchandise at the store, it's really helpful to tell the biometrics vendor the truth if you don't want their solution. The reason for this is that a good biometrics vendor will listen to your valid complaint, take the product back to his or her company, and give the design department a heads-up. ("See? I *told* you they wouldn't like the fingerprint readers when we shaped them like lizards' mouths!")

Although your organization may have done a try-it-before-you-buy-it evaluation, if you're doing a large-scale implementation, we strongly suggest that you include a pilot rollout as part of the plan. The larger your project is — by whatever measure, but especially by the number of users affected — the more important a pilot will be for your biometric implementation. As we discuss earlier in this chapter, a biometrics project has more ways it can fail than do most other IT projects — so going slowly will help establish learning on all levels: integrating with existing systems, installing, training users, and gaining user acceptance.

A *pilot* is a small-scale implementation used to validate integration, installation, provisioning, training — and whatever other steps are in the implementation plan — on a small portion of the full environment. For example, if you're planning on installing biometric palm-scanning readers on all the wiring-closet doors in all the buildings in town, we suggest you install one of those gizmos on just one door first, and let it "soak," "burn in," or whatever your favorite phrase is for "letting it run for a while to see how well it goes before we commit to the whole enchilada" (We don't, however, recommend letting your enchiladas soak; that will make them soggy.)

Before starting the pilot, you need to establish success criteria. By this we mean, when the pilot is over, how will you know whether it was successful? Your criteria needs to be objective and measurable, and the techniques for measurement established up front.

What will you measure? Be sure to measure the users. More than with most other types of projects, user acceptance and opinion are vital to biometrics — unless you enjoy the users coming in the night with torches and pitchforks, demanding that you fix the system or throw it out.

 We suggest you have your legal department people insert a clause in the contract with your biometric solution provider that makes the solution provider at least partially responsible for the outcome of the pilot. With more skin in the game, the vendor is more likely to help, and the pilot is more likely to succeed.

After the pilot, you're ready to implement the full solution, right? Don't be so anxious. After you've run a pilot, you have some nice, straightforward preparatory steps to take:

- ✔ Collect measurements
- ✔ Collect user-acceptance data
- ✔ Call a debriefing session where project participants, stakeholders, and even some users discuss what went well, what didn't go so well, and what changes should take place.

✔ Decide whether the pilot should be continued, started over, or whether the full implementation should take place.

✔ Obtain executive approval for decisions made during the debriefing.

Only after all this happens can you really be ready to throw the switch.

Throwing the switch

In the movie, *Young Frankenstein*, Dr. Frankenstein (or should we say, Frahnkensteen?) calls down to Igor, "Throw the third switch!" to which Igor warns, "Not the *third* switch!!" Frankenstein replies, "Throw it! *Throw it*, I say!!"

In real life, we really *do* feel the tension when we throw the third switch on a project — the switch that gives the new system life. As with Frankenstein's monster, the real complications of what you've just done might not be apparent for quite a while. For example, everything might work fine until some weekly janitorial-maintenance "event" happens, and you discover that these folks haven't been trained in the use of the new system, so the bathrooms are all out of soap and bathroom paper.

Phasing in a solution — if you can

If you're implementing biometrics on a large scale, we suggest you phase it in if possible. Implementing the system all at once is a little riskier. Here's why:

✔ If something goes wrong that wasn't seen in the evaluation or the test, there could be a lot *more* disruption as you scramble to fix it.

✔ Implementing all at once takes more resources (a *lot* more) and will probably cost you more as well.

✔ All-at-once implementations also have the unpleasant characteristic of being guaranteed to affect your boss either for good or for ill.

✔ Phased implementations allow you to choose who might be annoyed with you if things go poorly.

Creating backout plans

It's smart to anticipate that things can and do go wrong in technology projects. As you develop your implementation plans, you should think through each step; in particular, imagine how each step can be undone or reversed if something bad happens.

Here are some implementation principles that apply to any technology project — which we hope you'll apply to your biometrics implementation plan:

✔ **Create backups:** Back up systems, databases, configuration files — *whatever* you're changing.

✔ **Change the copy, not the original:** When a step in a process involves transforming data from an old format to a new format, preserve a copy of the old format so you can easily revert to it. If that isn't possible (for instance, if the dataset is just too large), then you need to develop the capability to reverse each transformation you do, in case something goes wrong and you have to go back.

✔ **Test each backout plan:** You wouldn't create an implementation plan without *testing* it — and it's just as important to test a backout plan to make sure it will work. (If you're skeptical right now, think about parachutes — and the reason for reserve chutes.)

✔ **Predetermine when to back out:** When things go wrong with a project like this, there's always a tendency to think "We almost have it. Just a little more time and we won't have to back out and restart." We've seen this tendency prevail until the CEO gets fed up and comes over to see who needs a kick in the pants. Know what the limits are beforehand — and stick to them.

✔ **Properly set expectations:** Let folks know how confident you are in the implementation plan. If you're working with really new technology (or you were unable to do a pilot), there's no shame in telling people that things might get bumpy — and that under some circumstances you're prepared to back out. This is *far* better than having to do a totally unexpected backout on installation day.

Informing users

The best technology projects include not only the plans, steps, and contingencies for making the technology work, but also everything needed to ensure that *all* users know the essentials: what's coming, why it's coming, how it will work, when it will be available, and how to use it.

This is like the old adage, *what if you threw a party and nobody came?* You wouldn't plan a party, order food and decorations, plan entertainment, and make other arrangements, but not tell anyone what, when, where, and why.

Equally poor form would be a last-minute e-mail to users (as shown in Figure 9-1), informing them of the big project just a few days before.

To: All employees
From: IT Helpdesk
Subject: Biometrics

On Monday morning, employees will be required to register their fingerprints at the first-floor helpdesk area. On Monday, everyone will be required to log in to the network using their fingerprints on the new fingerprint readers that will be distributed tomorrow.

This will make our systems more secure. Any questions, please call the helpdesk at ext. 1212.

IT Helpdesk

Figure 9-1: Notifying users of a new process at the last minute won't gain you any fans.

This approach of shoving a solution down the throats of all employees is short-sighted for several reasons:

- ✔ It doesn't solicit input or ideas.
- ✔ It makes employees feel that something is being done *to* them.
- ✔ It doesn't address potential concerns about privacy for the users' biometric information or how that information will be used.
- ✔ It doesn't take into account any employees who may be working remotely and won't be able to register from remote locations.
- ✔ It doesn't provide any sources for additional information.
- ✔ It doesn't address alternatives or workarounds.
- ✔ It doesn't list the names of the persons on the project team. (Is somebody afraid of the response?)
- ✔ It doesn't accommodate user training or special needs.

Methods

Even in a situation where a new biometric-based authentication or authorization solution is being implemented for use by all employees, in our opinion you still want to *market* it (in effect, "sell the idea") to your employees. Ideally, you'll be able to show everyone how the new system will have a positive impact on the organization, further organizational goals, and demonstrate a link between those goals and *personal benefit to employees*. For the project to be successful:

- ✔ Users must understand *why* the system is being implemented.
- ✔ Users must understand *how* to use the system.
- ✔ Users must understand the *impact* of the system on their work and on the company.

These three key objectives are accomplished with *information* that is *communicated* to employees in *meaningful* and *effective* ways. We break down this requirement as follows:

- ✔ **Information:** Employees need information that explains what's going on: what this biometrics system is, how will it work, and why it's being implemented.

  Be prepared to answer this worry: *Is it because they don't trust us?*

- ✔ **Communicated:** Different people assimilate information in a variety of ways. Some prefer on-screen text; others comprehend printed materials; spoken words work best for some; others want illustrations or diagrams. Information should be communicated over a variety of media, including (but not limited to) e-mail, voice mail, posters, flyers, Web sites, videos, and so on.

- ✔ **Meaningful:** The information conveyed must be meaningful. Users need to understand what is going on, how it will affect them, and why it is happening.

- ✔ **Effective:** If users are expected to do something, the information communicated must clearly spell out what. If there are rules, exceptions, questions, and so on, users must understand how it all works.

In a larger organization, these communications might be best handled by marketing professionals who know how to put together a message campaign that will reach all users in an effective, meaningful way. Larger organizations that lack these skills should consider outsourcing this task or prepare to face the struggles of doing it the hard way — either through more effort or with more active management, to make sure the users are "getting it" and understand what's going on.

Considerations

Let's face it: Even with an effective communication plan, some employees just never come out from under their rock. We don't know if they're just oblivious, or if they're so focused on their work that they never pay attention to any incoming communications, including the new mouse pad, screen savers and pocket protectors that scream, *"Biometrics are coming! Biometrics are coming!!"* (Okay, from a public-relations point of view, screaming mouse pads may not be such a good idea. But you see what we mean.) Even despite all the communications — the posters, the picnics, and everything else — some will be late to the party or won't come at all. We've had users respond to the PA announcement that there will be an outage related to the installation in a panic, with excuses ranging from "I never read e-mail from IT" to "Oh, you meant *this* Friday the 22nd."

Except in smaller organizations where you can put someone in charge of notifying each person personally, there will just be some who are left behind. You just need to be prepared to deal with a few people who will be *completely surprised* when they can't get into the building on Monday morning or can't log in to the network (*and what's that funny new mouse on my desk?*).

Training users

One great way to get users properly informed is to organize training for everyone. If the new system is really simple, then they may only need a 5-to-10-minute session; if it's more complicated, a little more time may be needed.

There are two approaches that can both be made to work: training before implementation, and training right afterward.

Pre-implementation training

You can train users before the biometric system is implemented. Training prior to implementation has its strong points:

- ✔ Training and marketing merge into one activity.
- ✔ You get users exposed to the new technology before it's actually implemented.
- ✔ Some users might ask questions on topics that the project team never considered.
- ✔ The training schedule can be a bit more relaxed, since they're not required to learn the system to get into work today.
- ✔ They'll know how to use the technology when it is implemented.

One disadvantage of pre-implementation training is that users are apt to forget what it was all about if too many days or weeks elapse between their training and their first opportunity to use the real system.

Post-implementation training

When we say "post-implementation" training, we're really saying that it's best to start training *just as the biometric system is being implemented* — we don't mean days or weeks afterwards. Also, post-implementation training doesn't mean that this is the first thing most users will hear about the new system. If anything, planning for post-implementation training means additional warnings and information about the upcoming events, since it will be an impactful event that must run very smoothly.

One really useful way to do training at implementation — or just after — is to combine training with registration, so each employee can learn about the new biometric technology and get fingerprints, palms, irises, retinas, or whatevers registered *during the training session.* When it's done right

- ✔ This approach lets each person register and then begin using the new system right away.

- ✔ People who aren't available right away (because of vacations, business trips, and so on) can attend the training later and still get registered.

- ✔ Often the mechanism for post-installation training and registration will need to remain in place for the life of the system, because new employees will need training and registration.

In an organization with more than a few hundred employees, you can hold training sessions all day for the first few days, and then less often. When you're registering users during training, hopefully you're keeping good records so you know who the stragglers are, and you can get them through the training to reach the 100-percent mark.

Dealing with user issues

The law of big numbers states that in a large enough company, there will be people who behave in almost any strange way you can imagine (those of you in big companies know exactly what we are talking about). Here are a few examples of the kinds of user issues you may have to deal with:

- ✔ **User doesn't show up at training:** Some just won't go to the training sessions. Maybe they're shy, too busy, or afraid of rooms with bright lights.

- ✔ **User refuses to be registered:** With privacy such a significant issue, you're going to have some employees who — out of fear or lack of information — will resist registering with the biometric system because they feel they're giving up valuable private information.

- ✔ **User refuses to use the system:** This one's similar to the last item: Some users will resist using the biometric system for a variety of reasons, including privacy or health (particularly with biometric systems that involve touching something like a fingerprint reader or palm scanner).

- ✔ **Users mess with the system:** . . . or with themselves, to "make a point." This could be related to privacy issues or some personal war with technology, but "accidentally" marking all their fingertips with indelible ink is an example we've actually seen.

Supporting Users

After a biometric system is deployed, it's going to require some attention. As with any electronics-based system, biometric devices are prone to failure — let's hope it's infrequent, even rare — but failures do occur. Not only the equipment, but users will need some attention as well: they need to know how to use the system properly, and this means knowing where to go for information in case they forget something (or if they're new).

Fault management

Every type of IT and electromechanical system is prone to failure; ask anyone who's ever used one over time. Whether frequent or rare, failures do happen.

When coupled with all its dependent systems, networks, and other components, a system as a whole will have a greater likelihood of failure than each of its separate components.

There are two ways that you'll learn of a failure of a component in your biometric solution:

- ✔ A monitoring system will alert you.
- ✔ A user will call you.

If you have a monitoring system, then both methods will apply in your organization. Without a monitoring system, your users *become* your monitoring system.

We suppose there's also a third option: *no one will tell you.* That would mean people are finding some way around your biometric system. In a building with multiple entrances, if you notice (say) weeds growing up around one of the back entrances, you might check to see whether the biometric device is working. If it isn't, then perhaps no one bothered to tell you — which is probably indicative of other problems beyond the scope of this book.

Monitoring

The beauty of a monitoring system is that you're more apt to discover a problem before your users are. Depending upon the type of biometric system you're using, you may be able to set up some monitoring of the biometric hardware (or the connected system) that will enable you to capture alerts and error messages and pipe them into a monitoring application.

Depending on the biometric system, monitoring may or may not be cost-effective. But remember: It's not just about the dollars, but about user productivity. Getting a truly meaningful and accurate estimate on any cost or productivity savings may well mean having a monitoring system that enables faulty devices to be repaired as soon as they're discovered.

Keep in mind, however, that a monitoring system is only as good as the functioning sensors currently in place — and dead systems sometimes tell no tales. A monitoring system will have a hard time detecting (for example) that a fingerprint sensor has a grime buildup that's about to start creating a bunch of false rejections until someone cleans it — but a heuristic monitor that spotted an increase in the FRR for the grimy sensor, prior to complete failure, might be helpful. If you rely on the systems themselves to report their health (or lack of health), keep in mind that sometimes components die without informing anyone (dead systems tell no tales). An independent monitor that polls each component from time to time will do a better job of letting you know when something is wrong.

Helpdesk

All but the very tiniest companies have an established protocol for whom to call when various systems in an organization (computers, lights, heat, water, and so on) require attention. In most organizations, a central help desk exists to accept complaints of every kind, which can be dispatched to the person or department that will actually fix whatever the broken thing is this time.

Organized help desks have references available that help them to figure out exactly what the caller is talking about when something is amiss. Before a biometrics system is implemented, the helpdesk needs to be equipped with several facts about the system so that they can help the caller.

Many companies have a goal of fixing problems during that first phone call. To accomplish this, helpdesk personnel have more extensive resources available that guide them through some simple troubleshooting (for openers, is the computer plugged in, is the power-strip light illuminated . . . ?). Sometimes helpdesk personnel have administrative access to systems so they can get an instant look at the status of a user's system or account — and can fix things easily while the user is still on the phone. Organizations with this "first-call resolution" as an objective will have to ensure that helpdesk personnel have the right tools and access — and train them on how the biometric system works so they can fix simple problems or help users to fix them on their own.

In the case if biometric authentication systems, keep in mind that they have something significant in common with other authentication systems: Typically they're set to lock users out after multiple unsuccessful attempts.

As with password-based authentication systems, the helpdesk will need the capability to reset a user's account to allow the user to log in. Helpdesk personnel should also be trained to ask useful questions about how the account got locked out in the first place — here the idea is to identify bad devices, stolen credentials (such as a username or badge associated with the biometric data), bad procedures, or improper use.

Repairs and replacement scenarios

The nature of electronics and electromechanical devices tells us that some devices will fail and require replacement. Your organization will need to be prepared to acquire (or stock) replacement parts and components — and be ready to use them if and when a biometric reader stops working for good.

Replacing a faulty unit is only part of the picture, however. Another aspect of the situation is to figure out how users access building spaces and computer systems when the biometric reader is broken. This, too, is a scenario that requires advance contingency planning. If a building-entrance biometric reader fails, how are you going to admit workers during a workday? (Propping the door open is probably a very bad idea.) Maybe you have to put personnel at the entrance to check IDs, or post a sign and have people use another entrance.

It's smart to fully document a device-failure-and-replacement scenario — complete with time estimates, a list of necessary resources (equipment, tools, and the people needed), and the expected impact of a device failure.

Publishing information for users

Most organizations prepare and publish information for users on a wide variety of topics — including benefits, compensation, policies and procedures, information systems, and so on. When a biometric system is introduced to an organization, the organization had better get some information prepared and made available so users have a handle on how to use the system.

Intranet

Intranets are the medium of choice for publishing nearly every form of in-house information. Information about your biometric systems can be included. Depending upon the type of biometric system you've implemented, there are a variety of things you can make available for users:

- ✔ Promotional materials that were developed for the initial implementation
- ✔ User instructions
- ✔ Troubleshooting steps (things to check if it's not working)
- ✔ How to register (or re-register)

✓ Original manufacturer instructions if they're useful

✓ Short video clips of users using the biometric system ("Oh, *that's* how I put my hand on the palm reader!")

✓ Whom to call for more help

One of the objectives for providing self-service information is to improve user productivity and reduce helpdesk calls (which are more costly).

Hard-copy materials

For the first few weeks or months, you may want to add hard-copy materials to your collection of information for users. One idea is to keep a supply of brochures or flyers that were used at the initial implementation. They can be given to new employees who join the organization after the biometric system is implemented — those materials can be tailored at the outset to be useful not only to employees who were around when the system is implemented, but also to new employees later on.

Printed materials, unlike your intranet, may find their way out of the confines of the company — and into the hands of people you don't trust. If you feel that any portion of the information in these materials could compromise the security of your system you should consider redacting that for printed materials.

Health issues

For the type of biometric systems that involve touch — especially those mounted on community devices such as building entrances — the topic of communicable diseases is sure to surface. Many common diseases are transmitted by touch, even indirect touch (consider doorknobs, telephones, and bathroom fixtures). Logic tells us that biometric devices such as fingerprint readers and palm scanners can — if conditions are right — serve as an infection vector if a contagious employee uses one of those devices. In most cases, the biometric device is neither more nor less capable of transmitting disease than a doorknob — which we don't usually provide additional sanitary measures for — but user perception is everything here.

In a situation where MSRA, influenza, or other diseases spread by touch are rampant, your organization may need to take steps to ease employees' fears and limit the transmission of communicable disease. Some possible remedies include these:

✓ Hand-sanitizer stations strategically placed just before or after a touch-based biometric station

✓ Dispensers of surface disinfectants used to clean biometric readers (make sure that the type of biometric reader is suited to such cleaners — that they won't become clogged with, or blinded by, cleaning agents)

We think it would be a good idea to develop some requirements around this issue. See Chapter 8 for more information on the methods used to select a suitable provider of biometrics solutions.

Maintaining a Biometric System

Biometric systems are generally thought of as a component in a larger information system. Most biometric devices contain microprocessors and firmware (software) that sometimes require updates and/or replacement. A well-designed system will allow these operations to be accomplished with little or no down time or retraining.

Software updates

A biometric system often includes software that runs on a workstation or server. Companies that produce software are usually under pressure to ship software before it's really ready — so those companies have to continue developing the software long after it's been put in the hands of customers. This means they must have a way to get software updates deployed.

We aren't going to tell you how to update the software in your biometric system — we presume that you have system administrators (or someone similarly harried) who know how to update software. As long as they follow good practices — such as performing backups before installing updates, and reading the release notes so the implications of the updates are fully known *before* installing the software — you'll probably be okay.

One measure that we do advocate for critical systems is that you *avoid combining them with other critical components on the same servers.* Not only is there potential for harmful interaction between the two (or more) critical systems in that situation, but also a kind of domino effect: An update that goes haywire could potentially affect multiple critical systems instead of just one.

Hardware updates

Biometric systems have a hardware component — a fingerprint scanner, palm scanner, iris scanner, earlobe scanner, whatever. Fortunately, hardware engineers seem to have better luck with releasing finished products than do software engineers. As a rule, few (if any) hardware upgrades are required, though they do happen from time to time.

Firmware updates

Many types of biometric devices have *firmware*, which is just software that's a part of electronic devices. Digital cameras, MP3 players, and DVD players have firmware, too.

Sometimes the manufacturer of a biometric product will update the firmware for its devices. As with software updates, you'll need to figure out how the firmware updates are actually performed, but you'll also want to read the release notes to make sure you actually *need* the firmware upgrade.

In some cases, firmware updates can be impossible to undo. More manufacturers are making firmware upgrades less dangerous, but there is still some danger that a firmware upgrade will turn an expensive piece of hardware into an oddly-shaped brick. That's what backup hardware is there for. (It *is* there, right?)

Change management

If you asked people in IT management if uptime and availability were important to them, most or all of them would affirm their importance. But if you asked them if they're using a change management process, fewer would say so.

Change management is the formal process of vetting every proposed change in a system prior to making the change. The steps in a change management process typically look like this:

1. **Proposed change:** Someone requests a change be made to a system. This change could be something as simple as a configuration change or as complicated as a software or operating system upgrade. The requested change should include these elements:

   - Description of the change

   - Business or operational justification for the change

   - Who will perform the change

   - When the change will be made

   - Impact of not doing the change

   - Risks associated with making the change

   - Backout plan in case the change is unsuccessful

   - Anticipated user impact (such as downtime while the change is made)

- Other systems affected by the change

- Test results (hopefully the change was tested on a test environment)

2. **Change review:** The proposed change is circulated for review among all the formal participants in the change-management process.

3. **Change approval:** Participants in the process discuss the proposed changes to identify any other risks or impacts. Then they can decide whether the change can take place as planned.

4. **Change wrap-up:** After the change is made, final recordkeeping can be filed to record the successful implementation of the change.

Organizations that use a process based on these steps experience far fewer unscheduled downtime incidents. When proposed changes are discussed among stakeholders, surprises (and their resulting downtime) have less chance to happen. In our experience, the most common reason for unscheduled downtime for a service is poor (or nonexistent) change management, not equipment failure.

Configuration management

In a computing environment, changes in configuration can become pretty complicated and hard to track when they take place on all the computers and in all the system layers (operating system, database, application, tools, and so on). Systems can get out of sync along the way; things can get really chaotic in less time than you'd probably think — especially in an environment that lacks adequate change-management processes.

Configuration management is the process of tracking all the configuration changes that occur in a given system or group of systems. In short, configuration management means keeping a journal of all the changes that occur in a system. The purpose of configuration management (or *CM*) is to provide a full history of the changes made to a system, to support any troubleshooting or diagnosis needed in the future.

For example, suppose a server supporting a biometric solution begins to malfunction: one of its processes has developed an apparent memory leak that causes the process to grow until it consumes all available memory — and crashes. With a good record of the changes happening on a system, it's not hard to determine that a recently installed operating-system patch is the culprit. With the cause of the problem identified, the organization can take steps to remedy the situation.

There are configuration-management tools available that automate the collection of CM information into one resource: a configuration-management database (CMDB). These tools are handy not only for tracking changes, but also for making automated changes on systems. In organizations with dozens or hundreds of servers, this is the only way to go.

The discipline of CM can be applied to a biometric system, even if there are no CM tools in place. It takes discipline, for sure — but good records kept from early on in a biometrics implementation — usually in the form of a journal — can be invaluable if problems crop up.

Upgrades

When you've had your biometrics system for more than a few years, your supplier is almost sure to tempt you with upgrades. The newer devices will be, um, *newer*, but also smaller, faster, easier to use, and possibly shinier.

Our advice to you is to keep your original devices for as long as possible. Your CFO (or whoever is responsible for the organization's finances) will appreciate your keeping the biometrics system going at least until the hardware investment is fully depreciated — which will probably be at least three years. Not only that, but upgrading to newer devices will involve labor, systems changes (maybe), and possibly user retraining as well. Remember that all expenses should be considered with respect to the *operational benefit to the organization*. If the upgrade doesn't provide a required benefit, then it's just about shininess. Get some sunglasses.

Chapter 10

Securing Biometrics Systems

• •

In This Chapter

▶ Identifying threats and vulnerabilities

▶ Investigating typical attacks

▶ Keeping biometric systems safe

▶ Employing security standards and practices

• •

A biometric system is, in the generic sense, just another information system. Like any information system, this one contains information — some of it sensitive — that must be protected from authorized disclosure, modification, or corruption.

Unlike most other information systems, a biometric system is used to protect other systems and assets. The value, then, of the biometric system is equal to the value of all the assets that it protects. More specifically, the value of a security failure associated with the biometric system is the sum of all the value associated with systems it is in place to protect. When you see it this way, you will begin to realize that the level of attention to security of a biometric system should be quite high.

This chapter looks at the vulnerabilities present in biometric systems and the threats that endanger them. We discuss typical attacks on biometric systems, steps you can take to strengthen the security of a biometric system, and the assets in the organization that such strengthening helps to protect.

Biometric System Threats and Vulnerabilities

Biometric technology has something in common with any technology used to protect valuable business assets: It will be relentlessly attacked until its weaknesses can be found and exploited. The biometric systems themselves may contain valuable information that becomes the target of attackers. Thus, in this section, we discuss threats to biometric systems and the vulnerabilities those threats target. It's important to understand those threats and vulnerabilities before we can hope to adequately protect our assets.

Before we go any further, let's look at the meaning of the terms *threat, vulnerability* and *risk*. Over the years we've found these terms to be used interchangeably and incorrectly. As with any industry jargon, these terms are tossed around and used by people who *do* fully understand their meaning, and by those who *think* they do — but don't really.

- ✔ **Vulnerability:** a weakness in a system that may permit an attacker to compromise it.

- ✔ **Threat:** a potential activity that would, if it occurred, harm a system.

- ✔ **Risk:** the potential negative impact if a harmful event were to occur.

The terms vulnerability, threat, and risk can be visualized like this: Imagine a game of chess, where one player has a very weak position, and the other player has a very strong position. The player with the weak position is unable to protect his king — this is a *vulnerability*. The weak player's king is vulnerable to attack – a position of high *risk*. The strong player has powerful pieces (such as a queen, bishops, and rooks) that are in low *risk* positions to easily capture the weak player's king — this is a *threat*.

And while we're at it, there are some other words we should discuss:

- ✔ **Attack:** the act of carrying out a threat with the intention of harming a system.

- ✔ **Exploit** (verb): the act of carrying out a threat against a specific vulnerability.

- ✔ **Exploit** (noun): a program, tool, or technique that can be used to attack a system.

Using the chess analogy again, the strong player could *attack* the weak player, *exploiting* his *vulnerability* to capture his king. The strong player's method of attack would be known as his *exploit* against the weak, high-*risk* player.

When you understand these terms, then discussions that use them will be clearer and more comprehensible.

Not all vulnerabilities are of the type that can be attacked or exploited by someone of malicious intent. Some vulnerabilities could lead to human error — whether of omission or commission — that could result in harm to a system. An example of this type of a vulnerability is a computer program whose user interface is so obscure that it leads to users who select the wrong options or perform the wrong tasks, resulting in errors and mistakes. Although an attacker may not have intent or means to attack this kind of a vulnerability, the vulnerability still exists — and can still result in trouble. Usually it's just a matter of time.

We hope that you're now clear on these important terms; when you are, the rest of this section will make a lot more sense. If you're a security expert, then you may have skipped this introduction, or perhaps it served as a handy refresher. Either way, here's where we get down to discussing real threats and vulnerabilities.

Natural threats

Natural threats to a biometric system are types of events caused by nature, also known as "acts of God." These include

- ✔ Flood
- ✔ Wildfire
- ✔ Earthquake
- ✔ Lightning
- ✔ Landslide or avalanche
- ✔ Volcano

We could include asteroid strikes, but you get the idea. All these and more have the potential to knock out biometric devices and systems. We don't want to turn this chapter into a miniature disaster-recovery book, but we do want to observe that many naturally occurring threats can damage biometric devices, systems, and equipment, as well as supporting infrastructure such as networks and servers. In many cases, an organization's disaster-recovery planning automatically includes support for such systems as biometric authentication servers. However, biometric devices are sometimes located in remote buildings that contain no other devices, while the servers keep chugging away in some other location, perhaps hundreds of miles away.

In some extreme cases, you should consider the effects of a natural disaster on the *availability* of a biometric system or measure. For example, people hunched over in a severe hailstorm and running for the door won't present good gait-based biometrics for the camera; any injury that alters appearance could throw off some of the other measurements; even an influenza pandemic could throw off voice-recognition systems.

Later in this chapter, the "Protecting Biometric Infrastructure" section discusses methods that can be used to protect biometric systems against natural threats.

Man-made threats

Man-made threats are those caused by actions — or inactions — of human beings, often resulting in malfunctions or damage to biometric equipment. Some of these threats include

- ✔ Fire and fire-extinguishing agents
- ✔ Malware (viruses, worms, Trojan horses, and so on)
- ✔ Sabotage and vandalism
- ✔ Hazardous materials accident
- ✔ Power-supply malfunction or failure
- ✔ Public-utility interruption
- ✔ Communications failure
- ✔ Social unrest and riots
- ✔ Terrorism and war

As with the natural threats listed earlier, this list is not meant to be comprehensive; rather, it should give you a better idea of the range of activities that can result in biometric systems not operating when they're needed.

Unlike natural threats, man-made threats may specifically target an organization for political, religious, ideological, or economic reasons. Some organizations just seem to attract ill will from certain individuals or groups. If your organization fits into this category, chances are you're already aware of this fact, and have taken measures to protect your business interests and assets. (If you haven't, now would be a better time to start than later.)

Biometric system vulnerabilities

Biometric systems are specialized; they include devices used to measure certain characteristics of individuals, but they may also include servers, operating systems, application software, and database-management systems. In this section, we discuss general vulnerabilities found in many types of information systems, as well as those specific biometric-related vulnerabilities. We break down the vulnerabilities by category.

Physical vulnerabilities

Physical vulnerabilities refer to weaknesses that may be present that would permit an attacker to physically access or harm a biometric system. Some of these vulnerabilities include:

- ✔ **Biometric device:** Although users need to be able to access biometric devices to use them, some device designs or installation methods may permit someone to damage — or even steal — a biometric device. It may also be possible for an intruder to alter the behavior of the device, enabling unauthorized persons to fool the biometric system into granting access. This physical vulnerability is exacerbated if the devices are deployed to control physical access to a secure area — and so *must* be placed in an unsecure area.

- ✔ **Cabling:** An intruder who can access cabling may be able to eavesdrop on the communications between biometric devices and servers, or between servers. This may enable an intruder to obtain sensitive information — and biometric data qualifies as sensitive. An intruder may also be able to block communications or alter communications through a *man-in-the-middle attack,* where someone who is literally in the middle of a communication between two parties can alter the communications going in both directions without the knowledge of either party.

- ✔ **Server:** If an intruder can gain physical access to a server, then he or she may be able to take complete control over it, whether by altering the software, the hardware, and stored data, or by stealing the server altogether.

- ✔ **Network devices:** An intruder who can gain physical access to network devices such as routers, hubs, switches, and firewalls may be able to eavesdrop on network communication, alter communications, or block communications.

Operating-system vulnerabilities

Biometric devices are often connected to a central server that contains the entire database of authorized biometric data — as well as the software that communicates with biometric devices. Servers all contain operating systems such as Windows or Linux which, if improperly managed, can have vulnerabilities such as these:

- ✔ Missing security patches

- ✔ Configuration errors

- ✔ Lack of, or improper, hardening

- ✔ Improper user access configuration

Database vulnerabilities

Your database-management systems (DBMSs) that store biometric data — such as Oracle, MySQL, and SQL Server — can have vulnerabilities of their own. They can permit an intruder to steal, alter, or destroy stored data. Some of these vulnerabilities in a DBMS include

- Missing security patches
- Configuration errors
- Lack of, or improper, hardening
- Clear-text (unencrypted) credentials stored on client systems
- Improper user access configuration

If you've noticed that the types of vulnerabilities between DBMSs and operating systems are similar, you're right! Both have many things in common — and suffer from many of the same types of security problems.

Biometric software vulnerabilities

In a corporate biometric authentication system, biometric devices are often connected to a central computer that runs a biometric software application that is used to manage enrollment and access requests. This biometric software may have vulnerabilities that could permit an intruder to steal data or unauthorized personnel to bypass biometric controls to gain entry into facilities or data. The operating system and database may be secure, but a vulnerability in the biometric software can just as easily result in a security breach.

Some of the types of vulnerabilities that could lead to these intrusions include:

- **Configuration errors:** The software could be configured in such a way that could result in security weaknesses including authentication bypass, exposure of sensitive data, or malfunction.

- **Installation errors:** If the biometric application software was installed incorrectly, this could result in exposed data, elevated permissions, or malfunctions.

- **Permissions errors:** If the biometric application's permissions are not properly configured, users of the application may be able to access more data or functions than they should.

- **Faulty logic:** If the biometric application's design has any flaws, it could be possible to cause the application to malfunction, resulting in elevated permissions for an intruder or access to sensitive data.

Many applications are Web-based; that is, users access the application via the web browser on their laptops, desktop computers, or mobile devices. Web applications have a host of specific vulnerabilities:

- **Injection vulnerabilities:** These vulnerabilities permit a user to insert computer code into data being sent to the application server. This can trick the server into executing instructions that expose or corrupt data.

- **Cross-site scripting (XSS) vulnerabilities:** This type of vulnerability results when an application accepts input data that is sent to a web browser without validating the data. Sometimes, as a result, a script executes on a user's browser that hijacks user sessions (among other kinds of trouble).

- **Parameter manipulation:** Web applications often use hidden parameters transmitted to and from a user's web browser. An application that does not examine the integrity of these variables may be manipulated to allow a malicious user to attack the application through this manipulation.

- **Session-management vulnerabilities:** Web applications establish unique sessions for its users; if the session data can be manipulated, this can result in a malicious user being able to hijack another user's session.

- **Cookie vulnerabilities:** Cookies are used to store small pieces of data on end user systems, usually for session management or user identification. If the data in a cookie is not well protected (say, encrypted), an attacker may be able to manipulate the cookie so it performs an authentication bypass or steals other users' sessions.

- **Buffer-overflow vulnerabilities:** Also known as a *buffer overrun*, this flaw exists when a program attempts to store more data than the size of a fixed-length buffer can handle. The overflowing data will corrupt other memory locations including code, causing program malfunction or even the execution of injected instructions.

- **Cross-site request forgery (CSRF) vulnerabilities:** These vulnerabilities are present in applications that do not perform sufficient authorization checks. A malicious user could construct URLs on his Web site that could, if clicked, result in the victim's browser performing an unauthorized transaction.

- **Malicious file execution:** Here an application (or a flaw in an application) permits a user to upload a file and have it executed on the server.

Biometric vulnerabilities

If all the preceding vulnerabilities weren't enough, there are some actions that can be taken in an attempt to fool or confuse a biometric application. Some of these of vulnerabilities include

✔ **Re-registration flaws:** An intruder pretending to be a legitimate user may be able to fool a biometric system into re-registering the intruder instead of the real user. For instance, suppose intruder Trudy is pretending to be user Ursula. When the biometric system refuses to admit Trudy, she explains, "It's not accepting my fingerprints anymore, I need to re-register them." If the system (typically aided by an attendant such as a security guard) permits Trudy to re-register Ursula's identity, then Trudy will have been able to use her own biometric to access Ursula's account. This type of attack often involves *social engineering* or some other stolen credential such as a password.

✔ **Matching flaws:** A biometric system with an excessive False Acceptance Rate (FAR) may permit unauthorized persons to access business premises or assets.

✔ **Replay vulnerability:** If an intruder can access cabling that transmits biometric information, he may be able to intercept and later transmit valid biometric data in an attempt to access information or facilities.

The vulnerabilities we're discussing in this section, if exploited, permit the attacker to break into whatever application the biometric system is protecting. If your online banking is protected by biometrics and an attacker successfully breaks the biometric system, the attacker is able to log in to your online banking application and do everything that you yourself can do if you logged in legitimately.

Attacks on Biometric Systems

Organizations deploy biometric systems to protect assets of value. As long as those assets *have* value, intruders are going to want to steal, corrupt, or destroy those assets, and they'll use any means at their disposal to do so. In the security profession, the techniques that an intruder might use are called *attacks*.

The purpose of an attack is to penetrate an organization, sidestep its controls, and (in turn) gain control of a desired asset. An attack might be direct or indirect, logical or physical, technology-based or human-based. Often attacks avoid using the same channels of access that legitimate users would use.

Attackers really have the upper hand when their attack strategies come up against our defenses. Here are some principles of attack and protection that make this point:

An attacker can . . .	*But we must . . .*
Attack the softest target.	Protect all targets.
Attack via the weakest path.	Protect all paths.
Exploit a chosen vulnerability.	Remove all vulnerabilities.
Attack at a time of his choosing.	Protect at all times.
Use a tool or technique of choice.	Protect against all tools and techniques.

Replay attacks

In a *replay attack*, an intruder has been able to record successful login sessions involving biometric systems or devices — and later tries to perform an authentication on his own by replaying the captured data.

For instance, if a voice-recognition biometric system is used to authenticate a legitimate user, let's say an intruder was able to intercept and record data that included the user's voice. Later, the intruder may attempt to access the same system (or facility) and will play back the recorded data captured earlier.

In the case of voice recognition, the attacker could record the legitimate user's voice over the air with a hidden microphone and later play it back through a speaker. Or the attacker could intercept a network transmission containing the user's voice (or the representation of the user's voice, in case it's encoded before transmission), and then play it back later by inserting it into the network.

Faked credentials

Some biometric systems may be vulnerable to attacks using faked credentials. Examples of such attacks include these:

- **Gummy fingerprints:** Several research articles have been published since the year 2000 that demonstrate that a few dollars' worth of ordinary chemicals can be used to lift a valid user's fingerprint and use it to manufacture a gelatin-based fingerprint that can even be placed over an attacker's finger. Such an attack could be difficult to detect — the gummy finger would be warm, like a real finger — and the attacker could literally eat the gummy finger once he or she had successfully fooled the biometric system.

✔ **Iris photographs:** High-quality photographs of a legitimate user could be used in an attempt to fool a biometric system. If a biometric system can't distinguish between a living iris and a high-quality photograph of same, then an intruder might penetrate a system or facility. In practice, this would require extraordinary resolution and reproduction as well as something to simulate the refraction of the outer layers of the eyeball, so it has never been done successfully.

✔ **Fake hands:** If an intruder can make a mold from a legitimate user's hand, he or she could possibly build a fake hand out of rubber, plastic, or some other substance in an attempt to fool a system that scans hands or palms. Nearly all hand and palm scanners have proof of life built in, but a fingerprint-only scanner could be more easily fooled in this way.

✔ **Facial photograph:** If a biometric system relies on facial recognition (as some laptop computers do today), an imposter may be able to fool such a system by holding a life-size photograph of the user in front of the camera. Surprisingly, this method has been shown to work with some of the less sophisticated facial-imaging systems. But even good systems are sometimes fooled by twins.

✔ **Recorded voice:** If an intruder can make a high-quality recording of a legitimate user's voice, he may be able to use it to successfully imitate the user by playing it back. This is possibly the easiest of these examples to attempt but also one of the easiest to protect against: Just have the user read a random phrase instead of the same thing each time.

Stolen credentials

Although grisly scenarios may not occur frequently, we've heard news stories of intruders who have actually stolen body parts from legitimate users to fool a biometric system.

In one case, as reported by the BBC, members of a gang in Malaysia chopped off a car owner's finger to get past the fingerprint-security system for the owner's Mercedes S-class automobile. And we have heard of a case where an intruder used a severed finger that was warmed by a flashlight.

Some biometric systems are more vulnerable to stolen credential attacks than others. Fingerprint readers are the most often exploited, followed by hand scanners. But the thought of a stolen eye is just gruesome to us — and unlikely to work, since most eye-based systems require proof of life (although criminals aren't likely to know that). However, we think you're pretty safe if your biometric system is voice-based — we have yet to hear of a stolen larynx being used to successfully fool a biometric system. And if your biometric system is gait-based, an intruder would have to steal *your entire body* — and then operate it exactly as you do! We think you're pretty safe if you use gait biometrics.

Bypass attacks

Intruders use every trick in the book to try penetrating a system or facility if it contains assets of sufficient value. If a biometric system is strong enough to resist a direct attack, an intruder will attempt to go around the system altogether — a bypass attack. Although this is technically not an attack *on* a biometric system, it is an attack on the controls protecting an asset.

One classic example of a bypass attack that we've heard about is a high-security facility that used biometric controls for entry. But the facility was located near a public road. An intruder could stop his car, walk to the rear of the building, and enter through a door that was frequently propped open so smokers could easily come and go during their smoking breaks.

Another bypass attack we're familiar with happened when an intruder got into a control room for a certain electrical power station that was protected with a strong steel door. The door could only be opened using a palm-vein reader. But the control room was full of hot, humming electrical equipment — so a flimsy steel vent wide enough for human passage was installed in the bottom of the door. *Mission Impossible* it ain't. But it worked.

Re-enrollment attacks

An intruder who is refused access to an asset by a biometric may be able to con an attendant, receptionist, or guard into re-enrolling the biometrics of a legitimate user — using the intruder's biometrics. Such a ploy wouldn't be too difficult to accomplish if the intruder had a fake ID and was a good actor or actress; often, the attacker is aided by information gathered from social-networking sites that describe the person being imitated, and even explain when he or she is scheduled to be on vacation (talk about a ready-made time for an attack . . .). We have yet to hear of someone successfully trying this approach, aside from authorized penetration tests, but we mention it here so you'll think out of the box (as we try to do) and consider all the possible ways past the controls and processes that protect a valuable asset.

System attacks

Someone hell-bent on accessing a protected asset may choose to attack a computer system that supports a biometric application. Several tools and methods can be used to attack a system:

✔ **Port scan:** In this case, an intruder has a program that sends data packets to a target system using every possible TCP/IP port number, in an attempt to discover the existence of a running service that could be exploited.

✔ **Password guessing:** In this attack, an intruder attempts to log in to the server over the network (or using an attached keyboard, if the attacker has physical access to the target system) — and tries commonly used passwords. If the intruder can log in, he or she may be able to try more attacks — such as application or database attacks.

✔ **Buffer overflow:** Here an intruder starts by communicating with a system tool or program, and attempts to cause a buffer-overflow error that could result in the intruder gaining control of the entire system.

✔ **Denial of service:** This is an attack where the intruder sends a huge volume of network traffic — or specially crafted messages — to the target system in an attempt to cause it to malfunction, rendering it unavailable for normal use.

Network attacks

An intruder may choose to attack a biometric system by attacking reachable network devices such as switches, hubs, routers, and firewalls. These network devices are not too different from servers or workstations: They are hardware machines with an operating system and other software. In the case of a network device, the software is designed for the special purpose of processing network traffic.

Some of the tools and types of attacks on network devices are as follows:

✔ **Port scan:** Here an intruder sends packets to a target network device, using every possible TCP/IP port number in an attempt to discover the existence of some feature in the device that could be exploited.

✔ **Password guessing:** An intruder tries to log in to the network device by guessing likely passwords. If successful, the attacker can control not only the device but also the network traffic flowing into and out of it.

✔ **Buffer overflow:** The intruder attempts to make the network device malfunction by sending packets specially crafted to cause a buffer-overflow error, which could result in the intruder controlling the device.

✔ **Network mirroring:** The intruder diverts the traffic of a target network and can collect network traffic as it occurs, gathering information that could be used to further compromise systems on the network.

✔ **Denial of service:** Here the intruder sends a large volume of traffic, or specially crafted messages, to the network device in an attempt to cause it to malfunction.

A successful network attack may permit an intruder to get closer to the goal of accessing or controlling an asset. A successfully hacked network device may be a stepping stone to an attack on biometric data (a valuable resource for further exploits) or the assets that the biometric system is supposed to protect.

Application attacks

An intruder can choose to attack an application that is used to control or manage biometrics, or another application that may be related or near to the protected asset. Remember, if an attacker can successfully attack *any* application in an environment, then he or she is one step closer to their ultimate target.

Techniques that an intruder can use to attack a system include:

- **Password guessing:** An intruder might guess user IDs and password combinations if he feels he might get lucky.

- **Buffer overflow:** The intruder could send specially crafted inputs to the application in an attempt to cause it to malfunction, including (possibly) the execution of injected code.

- **Input validation:** The attacker can attempt to confuse the application by sending it unreasonable information such as long strings of data or control characters. If the application doesn't validate the input data, it could attempt to process the data anyway, resulting in a malfunction.

- **Script injection:** The attacker can inject script commands into input fields in an attempt to alter or steal data from the application's database.

- **Cross-site scripting:** An attacker can attempt to implant malicious code in a Web page that executes on a victim's browser with a wide variety of possible results — including session stealing and execution of malicious code on the victim's computer.

- **Cross-site request forgery:** An attacker can place a link on his own Web site that causes the victim's browser to access a different Web site (where the victim has already been authenticated, or so the attacker hopes)— and, once there, to perform some action on behalf of the attacker. For instance, suppose attacker Alfred creates a URL on his Web site:

```
<img src=http://www.victimbank.com/transferfunds?account=Alfred&amount
        =10000>
```

If victim Mary clicks this image, and if Mary was already authenticated into www.victimbank.com, then VictimBank.com would transfer ten thousand dollars to Alfred's account.

Social engineering

Clever and resourceful intruders know that the easiest way to reach a target is by the path of least resistance. If an intruder is unable to successfully penetrate the technical defenses of a system or facility, he may instead rely on some unwitting employee to help the intruder gain access. Some examples of social engineering include:

- **Tailgating:** If an intruder is unable to enter a facility on his own, he can pretend to be an employee who has lost his key card (or index finger or eye) and follow an employee through a secured door. It's especially effective if the intruder is carrying some heavy object (say, a computer monitor or box of books); an employee is more apt to help the intruder into the building.

- **Remote access:** A clever intruder can make a series of phone calls to various people inside an organization to get all the pieces necessary to log in successfully to the corporate network. He can get the VPN URL from one employee, a username from another, and get a password reset from the helpdesk if they don't sufficiently validate the identity of the "user."

- **Loading dock entry:** Many reasonably secure facilities have a blind-spot when it comes to the loading dock. A good social engineer in a brown shirt and pants can often just walk in the back door with nothing but a clipboard.

- **Road apple:** The attacker leaves a removable medium lying around somewhere, in hopes it will get picked up — say, near the door to the lobby. A curious employee picks up the gizmo, takes it inside, and plugs it in — big mistake — whereupon it autoexecutes a Trojan or virus, granting the attacker access. This is especially effective if the medium is of some practical value — say, a USB stick or an SD card.

- **Dumpster diving:** Intruders can go through an organization's trash in the hopes of finding discarded printouts, memos, and documents that contain enough information that they can con their way into a system or facility.

Protecting Biometric Infrastructure

Biometric infrastructure typically consists of biometric readers, systems (servers, user workstations, or both), networks, and software. You have to take steps to protect all these components, one by each, using techniques discussed in this section.

This section will give you a head start, but you shouldn't consider this chapter an exhaustive treatise on protecting information systems. Instead, call it a starting point. You'll want to consult other texts for complete information on protecting servers and networks that use biometrics. We've included a list of books and other sources of the needed information at the end of this chapter.

Security-protection concepts

Information-security professionals tend to conceptualize or abstract the means used to protect information and information assets. It can be hard enough to understand the abstract nature of information systems (biometrics included); understanding how to *protect* them can, at times, be even more difficult.

Basic defenses

At the most basic level, biometric systems need to be protected in three ways: through confidentiality, integrity, and availability. Security professionals use the term CIA to denote these terms. Here's a closer look at these concepts:

- **Confidentiality:** Information must be protected from viewing by unauthorized parties and systems. Although in many cases the biometric system is serving as part of the means to protect information in the organization, the biometric-related information *itself* must be protected from onlookers.

- **Integrity:** The integrity of biometric-related devices, systems, and information must be maintained. All the components in a biometric system must be protected from unauthorized tampering. This includes the biometric devices themselves, as well as the systems and software that make it all work. Any unauthorized modifications to a biometric system may render it ineffective.

- **Availability:** The biometric system must be available for use at all times. If some condition or event makes the biometric system unavailable, then the assets that the biometric system protects may themselves be un-available when they're needed — which could disrupt business operations.

Next we need to talk about *controls*, which are the mechanisms that protect systems and information. There are four basic types of controls:

- **Detective controls:** These are controls that — you guessed it — *detect* an activity or event. Examples of detective controls are audit logs, event logs, and video surveillance systems.

✔ **Preventive controls:** These are controls that prevent an activity or event from occurring. A preventive control may include a means of identifying a subject, thus preventing unauthorized access. Examples of preventive controls include key-card entrance controls, data encryption, user ID or password login, smart-card login, and — oh, yes — biometrics.

✔ **Deterrent controls:** A deterrent control is one that is intended to discourage a subject from attempting some unwanted or unauthorized activity. Examples of deterrent controls include *No Trespassing* signs, video surveillance cameras and monitors in prominent view, and that really sharp-looking razor wire along the top of chain-link fencing.

✔ **Administrative controls:** These controls are the policies and procedures that define acceptable behavior from the organization's people — and the required characteristics of information systems. Examples include security policies, requirements, standards, and procedures.

Security purists will argue (correctly) that there are other types of controls as well. This isn't an exhaustive book about IT controls or internal audit, so it's not necessary for us to delve into these other types of controls other than to name them:

✔ Mitigating controls

✔ Automatic controls

✔ Manual controls

✔ Compensating controls

No matter what the type, however, sometimes controls fail. So here's a look at two control-failure modes. As you might guess, controls are really just abstract terms that denote the actual mechanisms that protect information and assets in some way. Sometimes these mechanisms can fail; when they do, they fail in one of two ways:

✔ **Fail open:** When a control *fails open*, this means that all events (authorized as well as unauthorized) are permitted. An example of a fail-open situation is the failure of a key card or biometric-controlled door buzzer, resulting in a door that opens with no more than a push. A fail-open situation puts protected assets at risk because they can be accessed by any party.

✔ **Fail closed:** In a *fail-closed* situation, all events are blocked, including those that should be allowed. An example of a fail-closed situation is a failure of a key card or biometric reader that prevents *everyone* from going through a protected doorway. A fail-closed situation disrupts business operations by preventing subjects from being able to access business assets or information needed to complete tasks.

Both fail-open and fail-closed situations are problematic; they put assets at risk and disrupt business operations. We can't say that all cases of fail-open are bad and that all cases of fail-closed are good — it really depends upon the specific control, what asset is protected by it, and what other controls still protect the asset.

Defense in depth

In the preceding subsection, we hint at the existence of more than one control protecting an asset. This concept is called *defense in depth*. This concept specifies that two or more controls, ideally of different types, work in combination to protect assets. Each control provides some type of protection by itself, and together they offer greater protection. Figure 10-1 shows a typical defense in depth. Examples of defense in depth, old and new, include these:

- ✔ **Castle:** The ancients understood defense in depth and got it right. A lot of treasure (or a beautiful princess) may be hidden in the innermost chambers of a castle that is protected by a moat, a moat monster (or possibly just a deterrent control in the form of a "Beware of moat monster" sign), a drawbridge, turrets for archers, high walls that are difficult to climb, inner courtyards with more gates, turrets, hostile terrain, and so forth.

- ✔ **E-commerce data:** An online merchant protects its valuable transaction data with firewalls, routers with ACLs, intrusion-detection systems, system-level access controls, database-level access controls, acceptable use policies, audit logging, and encryption. Notice that some of these controls are preventive; others are detective, deterrent, and administrative.

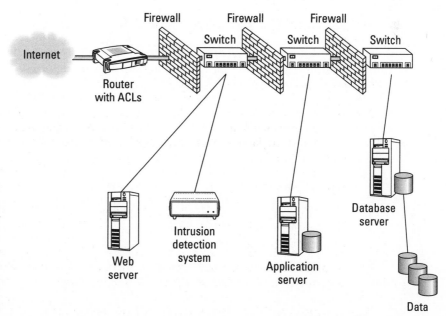

Figure 10-1: A defense-in-depth protection of an information system.

Protecting biometric devices

Biometric *devices* are the gadgets that perform the scanning or viewing in a biometric system. Recalling the concepts of CIA earlier in this chapter (confidentiality, integrity, and availability), we need to protect biometric devices by preventing the following antisocial activities:

✔ **Theft:** We want to keep someone from stealing the device, and for several reasons:

  • When stolen, the control will fail open or closed, disrupting business and threatening assets.

  • The thief may use the stolen device in an attempt to determine how others can be defeated.

  • He may steal the device, alter it, and attempt to replace it.

  • He may be short on cash and will try to sell it.

✔ **Sabotage:** We want to prevent someone from damaging a biometric device, which could prevent others from being able to use it.

✔ **Alteration:** We must be able to prevent someone from altering the device, which could lead to a fail-open situation that could leave assets unprotected; or to a fail-closed situation where no one can access business assets.

Protecting communications

Communications are present in most biometric systems, typically between biometric devices and servers. The nature of such communications should be considered confidential, and the security of the organization could be compromised if an intruder was able to eavesdrop on communications taking place in a biometrics system.

Several means are available for protecting biometric communications including:

✔ **Authentication:** A biometric system may authenticate the biometric traffic itself, as a way of distinguishing between authorized devices and imposters. Authentication might use MAC (hardware) addresses, encrypted handshakes (the electronic variety — hand-jive just doesn't do the job), or some other means.

✔ **Encryption:** Perhaps you can encrypt the entire communications channel between biometric devices and servers; if you can, then you may be able to prevent someone from being able to eavesdrop on your biometric system traffic and use it to launch a replay attack or a man-in-the-middle attack. If biometric communications are protected with encryption, then access to encryption keys must be restricted.

✔ **Cable protection:** You need to make sure that all cabling to biometric devices is protected from damage, sabotage, or alteration. This may mean running cables through conduits or raceways that are away from possible intrusions.

Protecting servers

Servers that contain biometric data or perform functions that support biometric systems must be protected from unauthorized access and manipulation. Such access could be harmful, could lead to the compromise of biometric data, or could allow unauthorized accesses to biometric-protected assets.

✔ **Security patches:** All available security patches should be installed on servers containing critical data (or that *connect to* critical data or functions). Failing to install security patches could result in unauthorized access when a hacker exploits known vulnerabilities.

✔ **Network access:** Only authorized, network-based accesses should be allowed. All unnecessary points of network-based access should be disabled, removed, or fitted with access controls. Putting network ports in the lobby (no! no! what where they *thinking?*) is a typical violation of this control.

✔ **Login credentials:** Valid login access to servers must be kept to the smallest number required for normal operations. Many intrusions into servers happen through guessed or cracked passwords for legitimate user accounts.

✔ **Hardening:** Servers should be "hardened" to protect them from unauthorized access, tampering, and attacks. Hardening servers includes removing unnecessary services and functions, enforcing strong login credentials, and using the strongest possible security configurations.

You have to provide adequate protection for all the servers in an enterprise environment — not just those containing sensitive business information or biometric data, but *all* of them. An intruder can easily find a weak server in an environment and use it as a stepping stone to a successful attack — even on a hardened server — if given the chance. Thinking back to the castle example earlier, it's like having a moat and drawbridge in the front, but a broad path to a flimsy door in the back.

Protecting biometric applications

The tools and applications that are used to manage biometric data and functions need to be protected, to prevent unauthorized access to biometric data. A compromise of any biometric application could lead to the compromise of biometric data as well as access to any and all assets protected by biometric controls. Several measures need to be taken to protect biometric applications, including the following:

- **Secure configuration:** The biometric application must be configured so it's secure. Unnecessary access methods and features should be disabled or removed.

- **Strong credentials:** The login credentials required to access the biometric management application must be strong enough to repel attacks that try to guess user IDs and passwords. Every user must have unique login credentials — none should be shared.

- **Proper installation and maintenance:** Biometric application software must be properly installed and maintained so it's free of vulnerabilities that could be exploited by an intruder.

- **Adequate role-based access controls:** The roles assigned to individuals who manage biometric access controls must be established, allowing only the minimum accesses required by each individual to perform stated job duties and to maintain the segregation of duties.

Protecting biometric data

Biometric data in an access-management system may be a target of attack if an attacker believes that data is usable to gain unauthorized access to protected assets (whether through a replay attack or some other means). For this reason, biometric data itself must be protected from unauthorized access and disclosure.

- **Access controls:** Biometric data must be protected so that only authorized applications, as well as authorized persons, may access it.

- **Audit logging:** Logging must be set up at the application, database, and system levels.

  Be sure to log all accesses and changes to biometric data.

- **Databases:** In many cases, database-management systems (DBMSs) are used to manage access to biometric data. Controls at the DBMS level are needed to prevent unauthorized access, and to log all accesses and changes. Database-management systems must be hardened, no less than servers, with measures that include security patches, secure configurations, and strong access controls.

✔ **Backup media:** Biometric data has to be frequently backed up to other storage systems or media, in the event that data in primary storage is lost or damaged. The three primary reasons to back up data are

  • *Hardware failure:* In the event a disk drive, controller, or server fails, data can be corrupted or unreachable.

  • *Software failure:* Although rare, a software bug can corrupt data or make improper changes to it.

  • *Disaster:* Should a fire, flood, earthquake, or other event occur, computer and storage systems may be damaged beyond repair.

Finding Sources of Security Information

There is no longer a need to start from scratch when you're looking to protect your systems and data. There are several excellent sources that provide strong guidelines, including these:

✔ **ISO 17799 and ISO 27001:** These are the international standards for information security management. These standards define practices for security governance, management, processes, and controls. One disadvantage of these standards is that they are not free: The documents containing the standards can only be purchased, and they are moderately expensive (a few hundred U.S. dollars each). You can purchase these from www.iso.org.

✔ **NIST (National Institute of Standards and Technology):** The NIST Special Publications 800 series are a series of very useful publications sponsored by the U.S. federal government in the area of information security and standards. Some standards (such as 800-76) are specific to biometrics; others (such as 800-53) are exhaustive lists of specific security controls that government offices are audited to check. The NIST Web site is a big place, but the 800 series can be found at csrc.nist. gov/publications/PubsSPs.html.

✔ **PCI DSS (Payment Card Industry Data Security Standard):** This standard describes in detail the controls required to protect credit card data. Although PCI DSS is credit-card-centric, if you protect biometric authentication data in the same way PCI DSS would require for credit-card data, you'll have done much of what is needed to protect it well. Information is available from www.pcisecuritystandards.org.

✔ **CERT:** The Computer Emergency Response Team at Carnegie Mellon University has a wealth of information for security management, incident management, and system hardening. CERT is located at www. cert.org.

✔ **SANS Institute:** This organization has information on incident response, server hardening, application hardening, and training. SANS is located at `www.sans.org`.

✔ **Books:** There are several good titles available that will help you to better understand how to protect biometric data and biometric information systems, including these:

- *Windows Server 2003 Security Bible*, by Blair Rampling

- *Security Engineering: A Guide to Building Dependable Distributed Systems*, by Ross J. Anderson

- *MCSE: Windows Server 2003 Network Security Design Study Guide*, by Brian Reisman and Mitch Ruebush

- *Special Ops: Host and Network Security for Microsoft, UNIX, and Oracle*, by Erik Pace Birkholz and Stuart McClure

- *The Web Application Hacker's Handbook: Discovering and Exploiting Security Flaws*, by Dafydd Stuttard and Marcus Pinto

- *Practical UNIX Security*, by Simson Garfinkel and Gene Spafford

- *Firewalls and Internet Security, Second Edition* by William R. Cheswick, Steven M. Bellovin, and Aviel D. Rubin

- *Testing Web Security: Assessing the Security of Web Sites and Applications*, by Steven Splaine

- *Network Security Bible*, by Eric Cole, Ronald L. Krutz, and James Conley

Chapter 11

The Future of Biometrics

In This Chapter

▶ Making current technologies better

▶ Investigating what the future holds

This chapter discusses what we expect to see down the road in biometrics — and some of the implications of these new technologies for authentication. Neither of the authors have a working crystal ball, but both make a living anticipating technological change and positioning companies to meet those challenges, so we might actually get some of it right — but we will almost certainly be wrong in some cases. Read this chapter with that in mind, and you're less likely to be disappointed when laser holographic nose-hair analysis doesn't make it to the mainstream.

Improvements in Technologies

In its first few versions, almost any technology has rough edges, wears out too quickly, and generally isn't as good as it *will* be in a few years — after folks have had a chance to complain about its deficiencies and engineers have had a chance to laugh at the horrible mistakes of their predecessors. Biometrics systems are still relatively new as a widely used, technology-based authentication tool, and have really caught on only in the last 15 years or so. There are many ways we can expect them to improve over time.

In this section, we take a look at biometric technologies discussed in Part II of this book — and make some educated guesses about where those technologies go next.

Fingerprint and palm scan

Since fingerprint-based biometrics are one of the most mature and widely-used of the biometric technologies, it might seem likely that we would see the least innovation in this area, but that's only if you ignore the fact that fingerprints are so darn useful they just can't be ignored. Fingerprints are so well understood as a biometric technology that new innovations in fingerprint capture and recognition are still valuable to the innovator.

Getting fingerprints

As of now, most fingerprint capture for biometrics is a very deliberate process where the user places their finger or fingers or palm into the capture device or runs them across the capture device and an image is taken. New advances in photographic technology with higher resolutions than were previously available might enable fingerprint capture at a distance, so that literally waving your hand at the pickup or briefly adopting the universal "I surrender" pose for a moment will be enough to authenticate the user. On the flip side, combined with the knowledge that it's quite possible to create a working fake finger with a print from a print-image, we might have to start wearing gloves in public to prevent someone from getting our prints with a high-resolution camera and a telephoto lens.

At the opposite end of the scale for acquiring fingerprints is the idea that to use a computer, you typically have to touch both the keyboard and mouse. Innovations in scanning technology may allow systems to constantly monitor the fingerprints of the person actually using a computer or use it as s second factor in authenticating a user as they type in a password. For this purpose, it wouldn't even be necessary to scan all the keys in a keyboard, maybe just the vowels. Th ntrdr wld hv t tp lk ths t kp frm bng dtctd.

Our best guess for acquiring prints is that new technology — some gizmo that captures not only prints but also proof that the print is attached to a living human being — will dominate innovations in fingerprint technology. Prints are too easy to acquire via simple means (such as lifting them off drinking glasses and ATM touch-screens) to really trust them by themselves for secure authentication.

I don't know art, but I know what I like

Humans have been developing insanely sophisticated biometrics processors for centuries, but we keep them in our brains. Since the ability to recognize faces and facial expressions is important to us, we devote a fair amount of brain power to performing this task, and a four-year-old can do it better than any computer system on the planet. If you doubt this, take a look at the level of artistic skill required to render an animal (say a skunk) face versus a human face. If you get the colors and basic shape of the skunk right, it will look okay to most eyes (but not to another skunk). Only a skillful artist can produce a good-looking rendering of a human face, because we humans are really good at processing the detailed biometric information in faces and expressions.

Comparing prints

In the before-time (25 years ago), when computers didn't have fancy graphical user interfaces, the idea of analyzing a high-resolution fingerprint, identifying the minutiae, and then comparing it to a new print in real time for verification was impossible. Here we are 25 years later and not only do computers all have graphical user interfaces, some of the graphics and animation are just for fun, because we have way more computing power than we really know what to do with.

Since we have all that computing power available, why do we have to break each fingerprint down to minutiae and use those limited points of comparison? Instead, it should be possible to use all the geometry in each fingerprint and gain significantly more confidence in the comparison. In fact, a sufficiently detailed image comparison, using all the available visual information of a fingerprint, would probably be able to detect and reject most of the simple ways of duplicating a print and finger. The extra processing power will also come in handy as any of the more ad-hoc methods of acquiring prints become popular, since the system might then need to rotate the acquired image and generally manipulate it until it matches the reference print best.

Hand veins and ultrasonic holography

Hand veins have so many advantages as a biometric tool that it's guaranteed to be a place where technologists spend time improving the acquisition and analysis techniques. Because proof of life is inherent to the technique, hand vein biometrics is already in widespread use in medium- to high-security installations with significant investment in this technology. Since the only way a manufacturer can get you to buy the same item twice is to make the new version irresistibly better in some way, we can expect to see lots of work put into improving this technology so we can replace last year's model.

Getting the image

Hand veins are really a subset of a larger group of hand biometrics that includes looking *inside* your hand rather than just at the surface. These methods currently use either ultraviolet light in the case of hand veins or ultrasonic sound for ultrasonic holography.

As it turns out, medical science is always looking for a good new way to look inside our bodies to get high-resolution images of three dimensional structures. The techniques that don't use ionizing radiation like X-rays, such as magnetic resonance imaging (MRI) are good candidates for looking inside us for biometric purposes. (Not to be confused with certain aquatic mammals that recognize people by looking inside our skins with natural sonar. Those would be biometric porpoises.)

It's true the MRI technology is currently pretty expensive to use for just doing a biometric scan, but we are constantly developing new ways to look inside the human body. It may be hard to imagine something that costs a million dollars becoming cheap enough for this purpose. Not that long ago, however, a simple laser was confined to laboratories and research institutions. Today you can buy one at the pet store for playing with your cat.

Clearly, looking at veins, arteries, bones, and muscles in high-definition color-enhanced detail will tell us everything we ever wanted to know about the unique aspects of our hands, fingers, and wrists, producing a few gigabytes of data about those appendages in a couple of seconds.

Using all that extra information

When someone walks up to you and asks for a couple of dollars to go place a bet on double zero, you check to see if it's really your brother-in-law before you hand over the money. In addition to making sure it's his smiling face, you quickly do a few more tests to see if giving him the money is such a good idea. Depends on the bother-in-law, but you might also take a whiff to see if he's been drinking, check his pupil dilation to see how much he's been drinking, and look at his wrist to see if he's hocked his watch already.

When we have as much information as we could expect to get with an MRI of the inside of your hand, we can do a lot more than just check to see if it's your hand, and some of that is probably even relevant to authentication, authorization, and access controls. For example, we can see from dilation of your veins and pulse rate that you might be under some amount of stress. Do we really want to grant access to the nuclear reactor? Maybe in that case we require additional personnel to accompany you.

Conversely, maybe your heart rate is low, but your blood oxygen levels are also low. Should we allow you to initiate the start sequence for the 787 you're about to fly to Germany, or maybe call an ambulance?

The only thing we can be completely sure of is that when additional information is available, we will find a way to use it in a biometric AAA capacity as we discuss in Chapter 2.

Signature

Biometric signature recognition is a tricky subject with respect to future developments, since it's really a technological way of recognizing an archaic form of authentication. If the next generation tends to sign their name less and less, will that eventually invalidate the basis for the biometric measurement of signatures? If you sign your name only a few times per year, will it still have a repeatable unique characteristic that will be useful biometrically?

Signing your name

Today, about all biometric signature systems record is the actual image of the signature, some information about how you moved the pen or stylus to create it, and the pressure you placed on the signature pad as the signature was created. Although that may seem about all there is to know about signing your name, there are certainly a few more unique data points that could be collected to better validate the signature.

The pressure that your fingers put on the stylus itself, not just on the writing surface is likely to be unique to your writing, especially when combined with and compared to what part of which letter you were shaping at the time. It's similar to *stylus-pressure dynamics* (the pressure of the stylus on the pad), only related to squeeze pressure on the stylus itself.

Also, it's certainly possible to measure the speed of movement, orientation, and direction of the stylus not just while in contact with the writing pad, but continuously, using accelerometers in the stylus itself. Once again, it's likely that for a motion that we repeat as often as signing our name, all these motions are unique to the individual in context with all the other information being collected at the same time.

Signing off

The writing on the wall suggests that we are stuck with using signatures for quite a while, even though other biometrics are far more personal, hard to duplicate, and accurate. That said, the problems inherent to good signature biometrics and relatively complex analysis required to produce good results mean that not many folks will be doing a lot of work in this area in the future. Personally, Mike signs checks when there is no alternative (maybe 2 percent of the time), contracts because there is no accepted alternative, and home loan documents because that industry seems to still be living in 1965. We can imagine that the generation of iPods and cell phones will look at signatures the way we look at wax seals and signet rings.

Maximum acceleration

Accelerometers are finding their way into all kinds of electronics these days. Since they are able to detect both acceleration due to motion and acceleration due to gravity, you can put one in a cell phone so the phone knows quite literally which way is up, and adjusts its display accordingly. We've had them in cars for quite a while now to tell the car when to deploy air bags. The accelerometers that IBM started putting into laptops a while back were designed to detect a peculiar condition — a complete lack of acceleration, which can only happen when the laptop is in free-fall.

Retina scan

Retinal imaging is an interesting technology for future speculation because it already meets almost any standards for accuracy we might require. The near zero FAR and industry best FRR don't leave a lot of room for improvement there. Nevertheless, there are some distinct disadvantages to retinal scanning that we can consider for improvement.

Scanning distance

To some degree, the distance your eyeball can be from a retinal scanner is dependent on how much detail we need, and therefore how high the resolution of the scanning system is. To understand what this is about, imagine a fence with a knothole in it, which represents your pupil. On the other side of the fence is a mural that you're asked to describe in as much detail as you can. Okay, now that we have the setup, describe the mural as you can see it from 10 feet away from the knothole and fence, and then describe what you see with your eye pressed up against the hole. Basic physical laws dictate how much detail we can see through a small hole at specific distances and the closer we are, the more we can see.

To take the analogy a bit further, now imagine that instead of using your naked eye, you do the same experiment with a telescope. The telescope has more light gathering capability than your eye and a wider aperture, so it's able to see more detail of the small piece of mural from 10 feet through the knothole than before. Back in reality, we know that one of the reasons that retinal scanning is so accurate is the fantastic complexity of the capillaries in the human retina, and so maybe a really high-resolution scan from further away will eventually be possible and still yield highly accurate results. It would certainly be nice to authenticate by just reading a sign placed strategically in front of the scanner but at a normal reading distance from your eye, instead of the normal three inches or so used today.

Another reason we expect to see work done on increasing the distance speed and intrusiveness of today's retinal scanning systems is that, along with facial recognition, iris scanning, and a few other technologies, medium distance high-speed retinal scanning could be used to scan in public spaces to help identify fugitives or other persons of interest to law enforcement. Obviously, there are legal and privacy concerns regarding using an ID system in this way, but facial recognition is already in heavy use in similar situations, so we expect that a more accurate method would be popular.

Health benefits

Not very many things can alter the structure or appearance of the retinal capillary system, and nearly all of them are related to disease. There are obvious eye-related problems such as glaucoma and cataracts, but also malaria, leukemia, and lymphoma have an effect on the eyes that can be seen in changes to the retina. It's possible that the system that allows you access to work every morning might also alert you to minute changes in your retina that could signal the onset of disease. We talk about ethical and privacy issues in more detail in Chapter 3, but consider the implications of this idea for a moment. Do you want your employer to know that you might have contracted a disease, possibly even before your doctor knows? Could the changes in your retina be considered Electronic Patient Health Information (a legal term defined by the U.S. HIPAA medical privacy law) and therefore subject to special protection? Would you rather that the system just ignored the changes and left you blissfully unaware? Given that the medically significant changes mean that you will start failing your authentication attempts, is there really any choice in the matter?

Iris scan

Due to the involvement of John Daugman's patents, and the fact that all current implementations of iris scanning currently use the Daugman algorithms, advances in iris recognition are largely dependent on the continuing work of Dr. Daugman and his team at Cambridge (noted in Chapter 13) and their commercial partners at LG — but not entirely so. The National Institute of Standards and Technology (NIST) conducted a large scale open evaluations of iris recognition software in 2006, the Iris Challenge Evaluation (ICE). Others are actively developing new algorithms that may yield non-infringing patents and begin to diversify the field a bit more.

One could even argue that there's not much future work required in iris identification, since it boasts the lowest FAR and FRR of all the generally available technologies, but accuracy isn't everything. There are areas that the industry is clearly interested in and will show progress in the coming years, no matter who is working on them. Also, as with retina scans, iris scans are subject to minute changes over time that the biometric system will have to incorporate into its methodology.

Scanning distance

Unlike retinal scanning, which has obstacles of geometry that make distance scanning very difficult, irises are visible from a distance. Current work at a distance of as much as 10 feet shows great promise. At greater distances, it can be harder to acquire exactly the right angle needed for imaging the essentially flat surface of an iris, but we can expect to see a lot of effort put into achieving iris scanning at great distances to make the scanning less intrusive, and to scan without the subject's knowledge. Examples of using photographs taken without the idea of iris recognition later being used for that purpose show (as with Sharbat Gula, detailed in Chapter 13) that good high-resolution photography will drive iris recognition more than processing for the foreseeable future. But again, there is a chance that legal issues may result in case law stating that high-resolution photographs of a person's face constitute an invasion of privacy. Currently however, images taken in public places are fair game.

Processing iris information

It's hard to use the word *perfect* with respect to a man-made process, but the Daugman algorithms appear to be close enough that lots of trying isn't going to buy much in terms of efficacy. Studies done on very large installations involving hundreds of thousands of individuals all in the same database indicate that current systems are as accurate as we need. New work on speed, accuracy in poor conditions, and oblique capture angles will certainly all be interesting and fruitful. Much of the work done on processing iris information will be focused on finding sufficiently innovative ideas that the Daugman and LG patents do not come into play.

Proof of life

One current knock against iris scanning is that it can be relatively easy to fool, using an image of the required iris. Since nothing inherent to the iris image or capture technology (digital photography) can show that the image presented is a real eyeball, or that a living brain is attached to it, there is a lot of future work that will use related information to establish this information including the following:

- Pupil dilation based on changing lighting
- Stereoscopy to get a view of the clear optical parts of the eye in relation to the iris
- Command- or signal-based blinking to see that an eyelid is attached and working

Made you look

With all ocular imaging systems, one challenge is to position the eyeball in exactly the same way as the last time you imaged it. As it turns out, that isn't as hard as it sounds. We have very accurate positioning systems in the muscles around our eyes that allow us to move the eye to focus on specific objects or images. If we can get you to focus on an object or image that never moves with respect to the capture camera, your eye will be positioned in exactly the same way with respect to the camera each time. For the folks that are identifying irises at a distance without our knowledge, that means they need to come up with something you will look at on your way by while they photograph your eye. Intelligence services will be hiring advertising agencies to consult in this area if they want to succeed.

Facial image

Recognizing facial images holds a special place in biometrics because we are all so good at it as humans, and thus far it's fallen so far short of its apparent promise. It's one of many areas of natural expertise gained over time; the human brain is just a lot better at it than computers are. We've only recently been able to create supercomputers with comparable abilities (such as the chess-playing Deep Blue in the late '90s).

Unlike most of the other image-based biometric technologies that focus on as much detail as possible over a fairly small surface, facial images are large areas, and traditional approaches actually toss away much of the detail in search of larger characteristics that remain unique. In the case of facial geometry-based algorithms, only the placement of features is examined. Combined with the fact that we rarely fully obscure our faces, and facial recognition becomes a prime candidate for biometrics at a distance.

There have already been many interesting tests of mass use of facial recognition biometrics with decidedly mixed results. Looking at the past as a map to the future, we can expect more testing and implementation of facial recognition technology and better results as the technology gets better at extrapolating facial characteristics from images at the wrong angle and as with most of biometrics, using additional detail provided by ever higher resolution images to provide better accuracy.

Facial image recognition is one area where we will see significant advances in how we process the information that do not depend on significant strides in image quality. The reason for this is not that we won't see the image quality gains, it's that there are a number of applications of facial imaging that use existing camera infrastructure, such as ATM cameras and security video cameras that will not be replaced quickly. These cameras were specified and installed without computer-based biometrics in mind and replacement will come at considerable cost, so we can expect to see older less capable cameras providing images for use in facial biometrics for a long time.

Since we know we can't force the replacement of every bank and parking lot security camera every time we add a million pixels to the current high-end capture camera, there will continue to be significant work to improve how we process poor quality images.

Facial thermograph

Thermographic cameras are expensive, compared to most other image capturing technology, and unlike visible light cameras, the cost is not being driven down by a large consumer market. Facial thermography does have interesting advantages over other biometrics that will likely promote additional research including low-cost infrared cameras with better resolution.

CSI For Dummies

If you've watched any crime dramas on television in the last 10 years, you have watched as attractive young forensic experts take a 320-x-400 image from an ATM, and then blow up the reflected image in the car bumper from across the street to then have a recognizable picture of a car license plate at an oblique angle to the bumper and hidden from the direct view of the camera. We're here to tell you that most of what you see isn't just really hard or super advanced — it's impossible. The information captured by the camera, which is limited to the camera's resolution and the light gathering capabilities of the lens, is all the information you have to work with. When you process images to sharpen edges or otherwise enhance them, you're essentially adding information to make the image clearer, and the information you're adding is at best a good guess. Blowing up a small part of a low quality photo presents exactly the same number of pixels, it just makes them larger. We're not saying that it's impossible to get interesting information out of photos using image processing, just that you're stuck with what the camera actually got.

How are you?

If you've ever worked in an environment with live guards at the doors or in the lobby, you may have come to know the guards and the guards know you. Possibly the first thing anyone says to you at work each day in such an environment is "Hi Jane, how are you doing?" A good guard will pay attention to your answer, and probably detect your general state of well-being, no matter what you actually say. Here again, this is something we humans practice starting in the first months of our lives, and get pretty good at after decades of practice. Facial thermography has the potential to perform many of the same kinds of emotional and physical evaluations as a human guard would in these situations, because many conditions of agitation, sobriety, and general unease change the blood flow to our face in ways that could potentially be used to tell the system more than just if you are who you say you are.

Peek-a-boo

Likely the most significant advances we will see using thermographic facial recognition will be centered on identification in challenging conditions where subjects are actively avoiding detection and identification. Since thermography does not use visible light — it uses the infrared your face emits as its light source — it can detect thermal patterns through fake moustaches, heavy makeup, scarves, and other disguises, and in total darkness.

Ear recognition

So far, ear recognition has resembled facial recognition in both the method of capture and analysis, but is superior to facial in accuracy due to the ear's stability in form over time and larger amount of unique detail. This stability and remote recognition capability make ear recognition a better candidate than its cousin facial as a nonintrusive and remote biometric identification.

The fact that our ears don't change shape over time means that we can consider image-analysis techniques that won't work well for things like faces that change with each expression. So, although the capture of ear images will likely follow the new technology for facial images very closely, the way that we analyze ear images will diverge from facial over time. Very esoteric techniques that use the initial image of the ear to produce Gaussian force field representations and other methods that treat the data as an abstract to provide results that are less affected by lighting and background variances are being explored with some success.

From a practical viewpoint, the ear provides such a better-behaved biometric than faces that we will eventually leave behind our species bias towards faces as an identification and see much more use of ears in situations where both methods are possible. A downside of ear recognition biometrics is that many ears are hidden behind hair styles. Would the widespread use of ear biometrics put a tousle in our hairstyles by requiring ears to be visible at all times (or at least at work or at the ATM)?

Speech

Speaker recognition doesn't suffer from a lack of fidelity in the capture or storage of the human voice. In fact, even the simplest systems are capable of capturing every possible nuance of speech and saving it perfectly. The single largest problem speaker recognition must overcome in the coming years is that it's based on both anatomy *and* behavior; and for speech purposes, both of these change over time, with illness, and at the whim of the speaker.

Basic truth

Advances in speaker recognition accuracy will be based on the concept that there is a basic physical absolute that forms the basis of the speech of an individual and that absolute does not change when the speaker has a cold, or ages a dozen years or so. To some degree, we unscientifically know this to be true, since we recognize voices that we know well through all these kinds of changes. I have even managed to annoy old friends that call after being out of contact for many years by correctly identifying their voice over the phone (not the most sensitive of audio instruments) and ruining their "guess who this is" game.

Only when we are able to eliminate the transitory effects of disease, hydration, purposeful obfuscation, and the long term effects of aging will speaker recognition become a really useful tool for authentication and authorization.

Call-in

Identification using speech is another matter though. Remember, using biometrics for identification is a one-to-many process where we compare one new sample to many samples in the database. Intelligence services and law enforcement will be pushing the envelope on speaker identification using samples from bomb threats, harassing telephone calls, and other collected samples. Speaker identification under these circumstances has the same problems with the behavioral and environmental changes to voice, but due to the nature of the identification problem presented; interested parties have no choice but try to incrementally improve on technique with inherent limitations.

DNA

For a technology that's used far more often in law enforcement than anything except fingerprints, DNA has a lot of drawbacks as a biometric technology and precisely because it's so widely used, a lot of time will be spent trying to deal with these drawbacks.

Waiting for processing

Current methods for processing and comparing a DNA sample to a reference sample take hours in the best case and more typically days. This rules out DNA for anything that requires immediate feedback, including most authentication or authorization tasks. (If you think the latency of user ID and password authentication is slow, consider that DNA-based login would take all day — by the time you logged in, it would be time to go home.) Although DNA is exceedingly good at identification, we will see research pushing the processing speed problem to allow DNA-based biometrics to be used in a more timely way. Oddly enough, one of the ways of doing that is to ignore the human DNA in a sample and focus on the simple stuff that's also present in the sample — bacteria and other parasites that are easier and faster to sequence, but as a collection offer up a DNA profile unique to the individual.

Collecting cells

We've come a long way from whipping out a needle and collecting a blood sample each time we need a few cells, but bucal swabbing (the inside of the cheek) each time you want to enter the server room is also more than a little intrusive. New techniques that gather as few cells as possible possibly by just vacuuming them off the back of your hand or your head might be a bit more acceptable.

Catching up with eyeballs

Both iris scanning and retina scanning are more unique to individuals than DNA. DNA comparisons are typically done using four to six non-coding regions and currently cannot distinguish between identical twins. Both iris and retina-based biometrics can distinguish between identical twins. Significantly faster processing of DNA would make it possible to use more than six regions and significantly improve the uniqueness of a DNA profile.

Dealing with privacy

Public perception of DNA identification is that it captures genetic information that reveals more than just identification. Although the genetic material provided does in fact contain such information, most identification techniques specifically choose to use areas away from or between genes to avoid this problem.

The problem is, once you give someone some cells for DNA fingerprinting, you have to trust them to throw away both the cells and the extra data not needed to identify you. It's hard to imagine how technological advances might answer this problem, but it's a significant hurdle that DNA biometrics must pass before it can be more widely accepted.

Gait

Analysis of gait is getting a lot of research attention as it provides another biometric tool that promises to identify people at a distance, possibly without their knowledge. The field of biometrics is still relatively new, so it's hard to speculate accurately about where all this research will take the field — but this chapter is all about speculation. We want to talk about all the biometric types that we cover in Part II, so we'll give the crystal ball a little extra juice and see what we see.

One of the most fascinating biometric events Mike has ever seen don't involve humans as subjects at all. In college he had a zoologist friend he hiked with in the woods and swamps of the Pacific Northwest. This friend wore glasses to correct rather bad distance vision; Mike has always enjoyed perfect uncorrected vision, so he was skeptical the first few times his friend pointed to a remote, dark-colored speck and named its species and gender. In every case, the guy was correct. How was he was seeing so much detail that Mike could not? Only later, in a conversation with a cognitive psychologist, did Mike understand: Apparently the zoologist wasn't seeing any more detail about the black speck; he was unconsciously using clues about location, movement, weather conditions, and time of day to deduce what he was seeing.

In order for gait analysis to realize its potential, we will see scientists figuring out how to emulate that zoologist's knack, using all the movement dynamics from the gait of the walker, but factoring in other movement and timing clues to help zero in on a more accurate identification.

Typing characteristics

Although not yet widely used, keyboard-typing-characteristic biometrics are so useful (and so simple to capture) that we're likely to see new and innovative uses in the near future. Current technology is not designed to capture some of the more interesting aspects of our typing though, so that's one place we can expect to see interesting new directions for this technique.

In addition to the typing dynamics already discussed in this book, it seems likely that additional characteristics could be used to further identify an individual. A keyboard modified to register (say) how hard each finger hits each key, typing rhythm, and actual fingertip-contact area for each keystroke would certainly add unique and characteristic data to the typing profile.

Also relevant to the typing style of an individual is what we would call that person's *composition style*, for lack of a better term. For example, Mike wrote exactly two term papers on a typewriter before he gained access to computers — and never looked back. His typing style is heavily influenced by the fact that he knows for certain that he can delete characters by just hitting the Backspace key, and make whole words or phrases go away by selecting them with a mouse and hitting Delete. That in itself distinguishes him from his parents, but not from the majority of the population at this point. The specific way that Mike deletes, however — and what he tends to delete or re-structure — is cumulatively quite unique to him. At this level of analysis, typing structure starts to blend in with the next topic for future speculation, sentence structure and linguistic analysis.

Sentence structure and linguistic analysis

Semantic analysis as a biometric technique is new enough that there are few current references to it beyond law enforcement and historical document analysis. It certainly has a bright future, though, if our analysis continues to improve in accuracy. Linguistic analysis in many ways is the ultimate biometrics at a distance, since it can be used to identify the writer of text, given a sufficient sample of known authenticity and sufficient unknown text to compare it to.

Natural language processing — of which sentence-structure and linguistic analysis for biometric purposes is a small subset — is currently studied by law enforcement (for example, Milt Jones's accurate profile of the Unabomber). It's also used for historical rabble-rousing such as debating who "really" wrote Shakespeare's plays and Lewis Carroll's *Alice in Wonderland*.

As yet, little of these analyses appear to be conclusive, but here again we look at the abilities of humans in biometric processing to see what might be done eventually with computers. From the experience of writing several books with co-authors, Mike knows that his father can accurately tell which chapters (or even paragraphs) are his, and which ones he did not write. Someday, when you log in to a computer and type a few sentences, the machine will determine whether it's really you or someone who has successfully guessed your user ID and password. We think we can expect improved techniques in the analysis of text to yield useful identification information from sample prose in the near future.

Brainwave

We know our human brains are unique, and the various impulses that make up our thoughts would be fantastically unique if we could really capture them in a useful way. Unfortunately, the ways we can watch brain function are either fantastically expensive — such as functional Magnetic Resonance Imaging (fMRI) — or dangerous to do a lot of (Positron Emission Tomography) or not very detailed (electro-encephalography).

Okay, current techniques are both impractical from a cost standpoint and too generic in the way they map current brain activity. So none are likely to be built into a USB key any time soon. That's not to say that all brains react the same way to the same stimuli, but similar patterns will happen across many brains, based on the current levels of activity we can measure. Oddly enough, the only way we *could* use current technology to authenticate that you are who we think you are would be to show you a *picture* of a password that only you know, and see whether the proper sequence for recognition fires in your brain. At that point, why not just have you *type in* the password and save a few million dollars in software development and equipment costs?

If we were to take a wild guess at where brain-function-based biometrics will eventually go, we'd say it will be based on watching the unique neural activity involved when the subject remembers something specific. This would need to be a very high-resolution technology that still wouldn't require you to lie motionless in a large magnetic doughnut after leaving your jewelry and other metallic items in another room. Not exactly conducive to a routine login to a computer, is it? But perhaps technology like this will be used to determine whether you're permitted to enter a building.

Identifying by odor

Really good odor-identification systems should be possible. After all, we know bloodhounds can identify people easily — and perfectly uniquely — by smell alone. The problem is that most current techniques for identifying the chemicals floating around in the air can really identify only a single chemical at a time — and body odor is a complex miasma of chemicals that collectively create the odor of a person.

A fair amount of work is underway to resolve the acquisition problem in odor identification; so far capturing odors for that purpose is difficult. Sensing what's floating around in the air *is* important if you're going to detect explosives in airports and drugs in various situations, so we'd expect that problem to drive the development of ever-more-capable electronic noses — which should, in turn, yield better odor-based biometric capabilities. Seems it would be cheaper to just use bloodhounds at the employees' entrance to the building and let the bloodhound decide whether you should be permitted to enter. It's not an elegant solution, but . . .

In data security, simplicity often wins out over complexity.

New Objects to Measure

With all the interest in using biometrics to identify, authenticate, and authorize, there will certainly be new objects or characteristics discovered that provide sufficient uniqueness for each individual to result in a useful biometric technique.

Although it's easy to imagine new things that we might be able to measure, it's quite difficult to imagine all the security, social, and privacy implications of new techniques until they've been in play for a while. We're going to take a stab at not only predicting some new biometrics that we might measure, but also at imagining some of the implications of each new measurement. Here are some of our thoughts on what we might reasonably see in use before we get too far into the future.

Behavior-based

Behavior-based biometrics rely on an interesting aspect of human learning, which is that when we learn to do a physical task well, we tend to perform that task almost exactly the same way each time. The more we practice, the more closely we match previous behavior. Because of this human trait, almost anything we do a lot could be imagined as a behavioral biometric.

Game play

If you're over thirty, this may not be the biometric for you, since your video game-playing skills haven't been honed by thousands of hours of controller use. The generation that can run, kick, and jump with a game controller almost as well (maybe better in some cases) than IRL (*in real life* for us oldsters) will likely develop some distinctive game-playing characteristics — how the controller is held, characteristic movements of each control, that sort of thing — that uniquely identify the person to the controller.

At the very least, we can imagine using such information to control access to differently rated games in a household; the ten-year-old could be detected and locked out from operating the more violent games, or presented with an age-appropriate version of the same game. There's no reason this couldn't be used for other kinds of authentication, though it's hard to imagine using a few minutes of *Tetris* or *Meteor* as an entry test for the office.

Reading dynamics

There's a fair amount of work being done to understand how people read and what that physical process really is. For example, an experienced reader doesn't read each character, starting at the far left and only having "read" a word after processing each letter. So exactly what is happening there? It's still a research topic, but we appear to recognize the word *shapes* for common words and only slow down to get more detail when a word we don't recognize shows up.

While people are studying these things, they've also been coming up with ways to track our eye movements while we read — and record not only exactly where we're looking on-screen, but also the movement our eye made to get there. We can suppose that the way we read is sufficiently unique to differentiate us, given good measuring capabilities and sufficient text to read. As with many biometrics, this one could also be used to supplement other techniques to produce a strong final product. You can imagine reading a short phrase while a camera tracks your eye movement, compares the iris to its known sample, and does voice-print analysis at the same time. Fooling all three systems simultaneously would be quite a challenge.

Basing biometrics on physical properties

This is really a third kind of biometrics different from behavioral and even different from physiological biometrics in a way. Physical properties biometrics are about measuring physical properties associated with an individual, but not in the traditional way of imaging a body part. Strictly speaking, DNA would fit better into this category than into physiological.

Electrical field

Our nervous system operates using electrical impulses that jump from the axon of one cell to the dendrites of another, carrying bits of information to and from the big collection of nerve cells at the top that we call the brain. Monitoring detailed brain activity and using it biometrically can identify people — and a scan detailed enough to accomplish that feat might be considered a bit intrusive, because it reveals quite a lot about your mental state and processes. A less-intrusive technology might focus not so much on the brain as on the electrical activity occurring elsewhere in the body; it might be combined with other biometrics that have a kinematic (movement-based) and behavioral (rather than physical) basis to provide additional, differentiating data of excellent quality. For example, if we could watch the electrical impulses heading down your arms to your hands while you type, while at the same time capturing your typing-behavior biometrics, the combined data would almost certainly be more accurate than we could get from either technique separately.

One potential flaw in an approach that samples the body's electrical field is that it could conflict with some versions of *personal area networks* (PANs) that use your body as the network medium. Versions of these networks are starting to appear, regulating access to cars, PDAs, and other stuff you have to touch before you can use. It would be easy to mistake the operations of these PANs for some sort of sophisticated biometric system. But in fact there is little or no traditional authentication happening; in these cases, the physical key is in contact with the subject (you), and transmits its authenticating information through your body to the device. And you can bet that techniques for hacking into a person's body (electronically speaking) won't be far behind.

Skeletal structure

Details of our bones and joints are certain to produce enough uniqueness to be useful as a biometric tool, if only we could figure out a way to get a good look at them that didn't involve dangerous radiation (if done daily) or gigantic machines with the annoying habit of wiping out the information on our credit cards. Since this technology is more of a potential future capability than a reality just now, we can safely make some guesses about where such technology might come from.

When we're looking for new physical measures of the human body, we can count on medical research to provide some. In many ways, the goals of medical researchers are almost identical to those of biometrics. Both want to find safe, nonintrusive ways to get extreme detail about various aspects of the human body. New techniques that could capture a noninvasive, low-risk, high-resolution photo of the skeletal structure would provide vast amounts of biometric information for identifying a subject.

General kinematics

Although gait is very promising as a distance-based tool, it's really a subset of the kinds of movement that might be recorded and used to later identify people. Those are known as *kinematics*. As a source of biometric data, kinematics can actually be classified as both physical and behavioral. The behavioral aspects of how we move are most evident when people do things that they do a lot — say, walking, opening a door, or eating. In effect, the activities we perform often wear a "groove" in our brains so we perform those activities nearly identically each time. Gait is one of the first such movements used as a kinematic biometric. That's because it's the most-often-performed general activity other than sleeping and sitting (both of which are harder to capture in a useful way).

Even when we're not performing well-known tasks, the way our bodies move has to obey some very specific rules governed by bone length, muscle attachment, and other factors that make up the physical metrics of kinematics. Given a large amount of sample data — and sufficient processing and analysis facilities — it's certainly possible to imagine identifying a person biometrically by watching how his or her body performs a novel task. Imagine being asked to bounce a basketball for a few seconds to get authenticated — that's an example of the basic idea here. Subtler methods would probably just watch you walk into the building, using not only gait but door-opening and head movements to authenticate who you are.

Non-genetic chemical

Although people have serious objections to DNA-based biometrics, there are lots of things swirling around in our bodies that don't have any of our genetic material, but may in fact uniquely identify us.

Little creatures

Viruses, bacteria, and microbes all have genetic material of their own, and they're all more easily examined than human DNA. There is also potentially more than one way for our micro-passengers to help identify us.

One way would be to examine the collection of strains of various beasties with the idea that while such simple creatures don't have enough complexity or uniqueness (in most cases) to identify individuals, there certainly is a wealth of information about groups of those creatures — and the collection of them hanging out in your spit might be unique enough to help identify you.

Another approach would be to use the most genetically advanced of the various microscopic animals hanging out with you to identify specific lineages. Since for them you *are* the ecosystem, your little friends are all pretty closely related to each other. (They don't get out much, so they have only other family members available for procreation.)

We don't know enough about the various genetic traits of flora and fauna living in, on, and around our bodies to know how much unique information we can extract from them as a proxy for you. Preliminary work in this area, however, seems to suggest that it's a possible biometric measure.

Antibodies

Neither of the authors have enough medical background to discuss the operation of antibodies in the human immune system in any detail, and (okay) it's not really within the scope of this book. So if you have any medical background at all, we suggest you skip this paragraph unless you're up for a good laugh at our expense; our explanation will touch only on the aspects useful to us in biometrics.

The body's response to the introduction of a pathogen — say, the common cold or rhinovirus — is to eventually create antibodies that bind themselves to the specific strain of pathogen so the immune system can seek it out and destroy it. The process by which this happens is wonderfully complex — and almost entirely irrelevant to our discussion of biometrics.

The interesting thing for us is that the antibodies for all the various pathogens we've encountered and defeated in our lives are still hanging out in our bloodstream. Now, think back on every sniffle, flu-bug, and generally fever-inducing time of your life since childhood; try to imagine someone else having *exactly the same leftover antibodies* as you. Not likely.

Current technology is not able to classify all the various antibodies in your system in any reasonable amount of time, but advances in medical technology may well sort that part out. Even with that sorted out, we still have to get some blood to do the testing, which seems a bit intrusive (to say the least). We can't imagine computer-operated hypodermic needles built into keyboards just yet (or don't want to, anyway).

So why is this form of biometrics interesting at all, you ask? For two reasons:

- ✔ It's a cool idea, and we technologists never let mere user convenience or comfort get in the way of a cool idea.
- ✔ It promises to be an accurate biometric based on medical science that does not record or use DNA at all.

The ability to do blood-based comparisons without recording anything about the DNA of the subject has the potential of helping law enforcement collect *elimination samples* (samples that are collected from persons known to be associated with a crime scene, but who are not considered a suspect — in order to identify their samples in the evidence collected and eliminate them) from people who would otherwise balk at giving the government their blood for DNA profiling. Although we might want to help out with a criminal inves-tigation, we might draw the line at the government potentially knowing about a genetic predisposition to (say) road rage. This biometric technology would provide a way to compare blood samples in a useful way, without the road-rage thing coming back to us in court years later.

Other chemical analysis

If it's possible to identify people by body odor, it stands to reason that a number of things about the human body must be chemically unique to each individual. Odor originates on the body and is carried away by evaporating sweat, so chemical testing the composition of sweat would yield interesting biometric results, even if we leave out the epithelial (skin) cells that we're likely to get with the sweat collection.

Likewise, in what might be really the ultimate biometric measure, we might grab a small chunk of skin and shovel it into a really sensitive mass spectrom-eter that can tell us exactly what chemical elements are in the sample and (with a bit more work) exactly what amounts of each — would that result in a sufficiently unique signature to identify the subject. Or are we all so alike chemically that this wouldn't be useful?

The answer to that question is probably somewhere in the middle. We are what we eat, as well as what we breathe, and (to some degree) what we hang out in. Subtle differences in diet, air, and environment all influence the levels of many trace elements in our systems. Although we expect that this kind of analysis would generate interesting results — and potentially have enough uniqueness to be biometrically useful — to our knowledge nobody has stud-ied this method in any really useful way. Yet.

Part IV
The Part of Tens

The 5th Wave   By Rich Tennant

EXPERIMENTING WITH THE WIRELESS LASER BEAM NETWORK

"Okay — did you get that?"

In this part . . .

No Dummies book is complete without these short chapters that come to you chock-full of references and tips for success. Here you'll find outstanding references, including some that will surprise you and broaden your knowledge about biometrics. And even if biometrics is new to you, it's been around for a long time, and we share the mistakes and pitfalls that we and others have found before you — we include them here to make your journey a more pleasant and successful one.

Chapter 12

Ten Tools Used in Biometrics

In This Chapter

▶ Exploring new-fangled ideas, techniques, and devices

▶ Understanding "soft" tools: survey skills, diplomacy, and business knowledge

*O*kay, the word "tools" is pretty broad — and here we use the broadest definitions of the word to discuss technologies, techniques, and even ideas that you can use as tools in understanding and implementing biometrics. The goal of this list is to describe some of the ways you can select, plan, and implement a biometric solution.

Credit-Card Fingerprint Scanner

This tool we find interesting not so much for the specific task, as for the concept that you can get *swipe-style* print readers (which require that you slide your print over a bar, rather than press a scanner) embedded into something as small and thin as a credit card.

If you're just looking at putting together a new biometric authentication system for your organization using existing technology, you may want to just skip to the next section; this section is written more for folks who are interested in how a biometric authentication or authorization might enhance their company's product.

The concept of putting a fingerprint reader onto the stuff you might want to protect from unauthorized access (such as a credit, debit, or proximity card) has been around for a while. If you're interested in what's out there, do an Internet search for *biometric memory stick* and look at the increasingly long list of such devices on the market. The idea of a working fingerprint reader on a credit card — which has, after all, little access to power, almost no storage, and a rough life of getting carried around in people's wallets — should lead designers down a whole new path in their understanding of where such technology can be used.

As a useful biometric tool, the low-profile fingerprint reader with miniscule power requirements has to be one of the most interesting things we've seen in a while. To get a look at a page describing this tool (and a nice picture of it), go to

```
www.ameinfo.com/58236.html
```

Contactless Palm-Vein Imager

Fujitsu Laboratories has been doing a lot of work on palm-vein scanning, so we mention two tools they have developed . The first is a new palm-vein scanner that looks like a small hot plate — but instead of heating up a package of ramen noodles for lunch, this one accurately reads the complex patterns of veins in your palm from a few inches away.

The new style of reader solves the hygienic issues of having many people touch the same scanning device hundreds of times each day to be granted access to a facility. Since there is no required contact, no germs are transmitted from person to person. In health-care situations, this is a critical advance.

Also addressing health-care is the idea that since this is an infrared imager, it can "see" through latex gloves pretty well — and authenticate or identify workers without any impact on a sterile environment.

Generally, people feel more comfortable using a device that they don't actually have to touch. In addition — unlike the body-part-print-based technologies — you don't accidentally leave what's measured (your hand-vein patterns) on stray drinking glasses. You can be pretty confident that the source information for the biometric is secure. You can find a picture and more description of this technology at this link:

```
http://www.fujitsu.com/global/about/rd/200506palm-vein.
     html
```

Mouse-Embedded Palm-Vein Scanner

Also from Fujitsu, the mouse-embedded palm-vein scanner is really just a specific instance of the same technology as the contactless palm-vein imager discussed previously: To be authenticated, the user just holds a palm above the imaging device built into the mouse. In addition to this no-touch convenience, such tools are worth special mention because they accomplish something important: They don't take up more space on the desktop.

Folks now make computer mice with fingerprint readers, palm-vein readers, and even an iris-reading mouse. We feel that a new authentication technology for the desktop should have as little disruptive impact as possible — not only on users and their typical workflow, but on the increasingly cluttered desktop workspace. Embedding technologies like these into items already existing on the desktop means you don't have to find yet another nook for the new biometric-authentication tool. Another great example of this kind of design is USB hubs built into computer flat screens — now that we have a screen with a hub in it, we'll never buy another desktop hub.

To get a look at examples of each of these desktop authentication technologies, take a look at these links:

- **Fingerprint mouse**:

  ```
  http://gizmodo.com/gadgets/notag/microsofts-fall-keyboard-and-mouse-lineup-
          revealed-19707.php
  ```

- **Iris mouse**:

  ```
  http://www.gadgets-weblog.com/50226711/qritek_irisrecognizing_iribio_mouse.
          php
  ```

- **Palm-vein mouse:**

  ```
  http://www.ohgizmo.com/2008/01/08/ces-2008-fujitsu-palmsecure-vein-
          verification-mouse/
  ```

Cardio-Signature Reader

Aladdin Knowledge Systems has brought to market a new kind of biometric reader that is both a useful tool *and* the first of its kind, reading a new kind of biometric: the electrical signals that make your heart beat. Instead of focusing on the main signal (as you'd see on an EKG), however, these readers look at the more subtle "under-patterns" that lie beneath the main control signals — and appear to be completely unique to an individual, given enough capture detail.

The Aladdin system is one of a whole new set of biometric tools that are cropping up: *biometric signatures*. For the most part, these will be looking at subtle signals from autonomic systems in the body that produce enough unique information to identify individuals. We wouldn't be surprised if, at the appropriate level of detail, something like the muscle-twitch patterns in an eyelid before a blink will become a unique identifier (and might do dual duty as a lie detector).

You can get a look at the Aladdin cardio device at

```
www.securityinfowatch.com/online/News/Aladdin-Touts-New-Cardio-Signature-
        Biometrics-Authentication/7309SIW1
```

Laptop Facial Recognition

For quite a long time now, laptops have had the option of an embedded video camera — offering live video chat, videoconferencing, and maybe something more interesting than dull spreadsheets to hackers who get remote control of your laptop.

Lenovo — possibly among others — has found a new way to use that embedded camera: to authenticate users using facial-recognition biometrics. Several Lenovo models use a program called "VeriFace" to authenticate a user to the Microsoft Windows Vista operating system. Because this tool is integrated with Vista, all the legitimate user has to do to unlock a locked screen is to sit down in front of the computer and it will unlock. The whole package actually gets good reviews from people who own those particular Lenovo models. In addition, as with various biometric methods built into computer mice, this one doesn't consume any additional desktop space — especially important for a laptop since the "desk" might be, well, your lap.

The use of this tool isn't limited to just Lenovo laptops by any means. At least one other software company out there, Banana Security, has produced software for this purpose. Here are a couple of Web sites that offer more information:

✔ For more about the Lenovo laptop, visit

`http://blogs.consumerreports.org/electronics/2008/04/a-lenovo-laptop.html`

✔ For more about Banana Security's Keylemon software, go to

`www.keylemon.com/`

Biometric Timecard Systems

Some studies indicate that time theft can account for as much as ten percent of the total payroll in a company that pays hourly. A traditional approach to solving this problem is to tie an entry badge with a magnetic stripe on it to the timecard system, so you have to have your card handy to get into the facility, and you have to have it with you to clock in. Creative people have found dozens of ways of circumventing this kind of system, and it introduces a losable card that may also create an ongoing expense.

The new biometric tool for solving this problem is a biometric timecard system that identifies *and* clocks in users when they present their biometric information. Since time theft is a fairly widespread problem, there are a number of manufacturers making integrated biometric timecard systems that use fingerprint, hand-vein, and hand geometry to identify individual users and start clocking their hours.

One thing we find interesting about these systems is that they're sold as a package, with the biometrics and the time-logging system all built into a single device. Truth is, *any* good biometric system with a logging facility (a feature no *good* one would be without) would work just as well for time-tracking and check-in. The only change from a normal medium-security biometric installation would be an added requirement: that you check out when you leave.

Biometric Flash Drives, Portable Hard Drives, and Stuff

A fingerprint reader on a credit card — a useful design landmark we mention earlier in this chapter — isn't all that farfetched; some new tools are taking advantage of this concept. These devices do not, for the most part, provide tools for creating a biometric authentication system for your home or office. (Not yet, anyway.) But they can be used biometrically to secure devices and transactions.

Portable storage of all kinds is starting to get a bad rap from security folks — and for good reasons. One of the more serious problems we face in information security is the loss or theft of devices that contain huge numbers of private records. The authors haven't figured out why it's necessary to have a hundred thousand or so customer records on a keyfob or laptop that you leave in the trunk of your car while you watch a Little League game, but it happens. Such security lapses are one reason about 325 million customer records containing private data have been exposed over the last four years. One great way to reduce the impact that theft of portable storage can have on the organization — and the customers — is to integrate encryption into the device itself, using biometrics to keep the authentication quality high and avoid the lost-password problem.

Another security problem that this kind of tool promises to solve is securing transactions that happen in public places on public networks. We've known for a long time that such transactions need to be encrypted — at the very least, so prying eyes can't grab the credit-card number as it flies by. Another problem with these transactions is harder to solve — being sure who is initiating the transaction. Is it you, or is it someone in possession of your credit card and/or cell phone? Some early vending-machine and cell-phone integrations trusted whoever was in possession of the cell phone to buy a candy bar using that phone number. Some new phones and PDAs have built-in fingerprint readers that allow a vendor to verify securely — through the cellular, Wi-Fi, or NFC connection — that the individuals initiating transactions are who they say they are. In many cases, the biometric authentication goes no farther than the device — which already knows the biometric of the registered owner, and can do the verification locally.

The "tool" in this case the range of devices out there that already have good biometrics sensors installed — and that you may be able to use for your particular purpose if you get creative.

Survey Skills

This is the first of three "soft" tools that we feel are important to your ability to be effective in deploying new biometric technologies. One of the most important things that you will need to know when preparing for a new biometric technology is certain kinds of information about your users' physiology, behavior, and attitudes.

No biometric tool will be well accepted if it depends on physiology that is (for some reason) not accessible or usable in your user base. We know of an information-technology professional who designed a new fingerprint-identification system for accessing secure storage facilities for a utility district. He chose a great system, tested it on himself and lots of his office mates, and then started putting it in place for the storage facilities. In fact, enrollment of the workers was going very well, but over the course of the first day, he discovered the serious flaw in his thinking: Utility workers start the day with clean hands, so at that time, registration worked fine. As the day progressed, dirtier and dirtier hands were being used for authentication, which wasn't great for the equipment, and wasn't working so well for authentication, either. It turns out that a number of the workers saw this problem coming, but "nobody asked them" so they just watched the train wreck.

In large organizations, it's nearly impossible to anticipate the work conditions for every category and class of employee, so don't try. Instead, find good ways of asking them about those conditions — and listening to what they have to tell you. Even the most uninformed computer or technology users will know more about how their day rolls out than you do, and they can give you clues that will help get things right the first time. In the utility company's case, things went from bad to worse as they discovered that a significant number of the older workers' hands were so worn from manual labor that they were nearly unreadable.

Diplomacy . . .

. . . is a skill that everyone should cultivate — but information technology workers have even more need of it than others. That's because as an IT guru, you're constantly disrupting the workplace. It's not entirely your fault. As it turns out, information technology is disruptive by nature.

Biometrics projects change the way we gain access to the tools and locations we need to do our jobs — and that change can be on the extreme side of disruptive if it's not handled well. The transitions have to be smoothed. If unforeseen circumstances make the transition from the old way of doing things to the new biometric system *less* than smooth, you'd better know how to unruffle the ruffled feathers. Keep reminding yourself that no concern is "silly" just because it stems from ignorance of the system or misperceptions. Nobody can (or should) be expected to know the system the way you do — and if a particular concern seems silly, that just means you should be able to easily allay that concern with some well-chosen words.

We mention the problem of privacy often in Chapter 3, so we won't talk about the details here — except to say that many people you work with will have privacy concerns regarding *any* use of a biometric system, and you'll hit a snag in deploying biometrics if you're not prepared to help those folks understand the organization's point of view. Often, the best way to accomplish that is to understand *the users'* point of view, so they don't just dismiss your ideas.

Business Knowledge

There's one more important body of knowledge needed for a successful biometrics project: a deep understanding of the business processes into which you'll be integrating your biometric techniques. Even if you have an encyclopedic knowledge of the technical aspects of biometrics, the systems you will be installing onto, and the details of the installation process, you have to understand why the processes you're working with exist in the first place. Then you have a handle on the new biometrics technology will make that process better, more efficient, or maybe just compliant with laws or regulations. Remember, it's ultimately about the business!

The job of a good information technologist is to enable business practices, and it's no different for security people. Yes, your focus is to build security into those processes, but the processes themselves are what's important, not the security itself. The security you should be building is a product of business decisions about risk management, and not some absolute standard of what is secure and what is not.

A really great biometrics project will be one that helps the organization manage risk by improving authentication or identification techniques for processes that have been deemed to embody too much risk from the perspective of identity management. A bad biometrics project just arbitrarily makes identification management more secure, and stops there. It may not have addressed *managing risk* at all. Don't just spend money to "make security better."

Chapter 13

Ten Biometrics Web Sites

In This Chapter

▶ Privacy and biometrics

▶ Biometrics in the news

▶ Biometrics research

*T*here are quite a lot of biometrics Web sites and other sites out there with good biometrics information — so many that choosing just ten great ones is a bit of a challenge. What we've done here is not so much collect the ten absolute best sites we could find (that's a moving target), but rather we chose ten helpful sites from several categories to provide a wider view of biometrics sites on the Internet. The Internet changes daily, and in an emerging field such as biometrics, we can just about guarantee that by the time you read this, there will be more sites than we could have included — and some of the links we provide will be somewhat outdated. Use what we provide here as a starting point for your own exploration and see where it leads you.

National Geographic and Sharbat Gula

We mention this particular case in Chapter 3 in connection with biometrics in our daily lives — but it's worth highlighting the *National Geographic* Web site's pages that talk about the fascinating tale of a young Pashtun woman caught on camera by photographer Steve McCurry in 1985. Mr. McCurry and a team of biometrics scientists authenticated her 17 years later — using the iris images from the original 1985 photograph to compare to a more recent photo of her.

The story and the Web site are interesting to read from a general human-interest standpoint, but we also use it all the time as a biometrics Web resource to illustrate some key points:

✔ **Authentication is a far different process than identification.** In this case, the identification was accomplished through strong journalistic and detective work, on the ground in a place where public records are rarely of help and "Gula" is a very common name for women, meaning "flower" in the Pashto language. In contrast, the authentication of the identified woman — confirming she was the same as the original — was only slightly complicated by the fact that the original image *was not intended to be used as an iris biometric.*

✔ **Some of the best biometrics — such as iris and vein patterns — do not change over time.** The ability to perform a very high-probability match using a single 17-year-old sample collected poorly (from the biometric point of view) shows how well these kinds of measurements work over time.

✔ **We will see all the privacy issues surrounding biometrics only over time and with experience.** Sharbat Gula wasn't particularly happy with having her picture taken in the first place, according to her own account, and was unaware of very broad dissemination of her face on the cover of *National Geographic*. Her face changed greatly over the years, and most people would not have made the match with the original photo, but for her family's memories of the event. The fact that we can prove that it was this particular woman, placing her and her family at a specific time during a time of war in her homeland, certainly has privacy implications.

You can find the story about Sharbat Gula at

```
http://ngm.nationalgeographic.com/ngm/0204/feature0/
                    index.html
```

Electronic Frontier Foundation

The Electronic Frontier Foundation (EFF), a strong advocate for citizens' rights in the digital age, provides information, advocacy, and (in some cases) legal support to people whose electronic freedoms or privacy are at risk.

Not too surprisingly, the EFF has written a well-founded and well-reasoned paper expressing the EFF's concerns about the use of biometrics, which is posted here:

```
http://w2.eff.org/Privacy/Surveillance/biometrics/
```

In this paper, the EFF does a great job of describing its concerns, though the assertion that the accuracy of biometric systems are impossible to assess seems to indicate that this was written before some of the large-scale installations were in use, which are now providing high-accuracy data from large-scale use that is not tainted by manufacturers or vendors.

The discussion on "Current Biometrics Initiatives" is definitely worth a read if you want to understand what the U.S. federal government is up to in this area. This area, once again, makes us think that the article is a year or two old, as it doesn't mention the implications of the facial information captured by U.S. Department of Homeland Security's Real ID initiative (which we discuss further in Chapter 3).

Possibly the best section of this paper is "Attributes of Biometric Systems"; this section not only talks about biometric systems in the abstract, it also discusses all the most common types in some detail. It's a different perspective from that presented in this book, and well worth reading.

National Biometric Security Project

The National Biometric Security Project (NBSP, at `http://www.national biometric.org/`) was established as a nonprofit organization shortly after the events of 9/11/2001 to help the government and private industry protect U.S. national infrastructure through the use of biometric technologies. Keep in mind that this organization is primarily focused on protecting infrastructure, and all its material is provided relative to that goal, so it's really of marginal use if you're just interested in figuring out a better way to use the fingerprint reader on your laptop or installing a biometric lock on your house.

Unlike many of the initiatives undertaken immediately after the 9/11 attacks, the NBSP understood from the start that there were privacy issues related to the protective measures they would be assisting — and they spent some time and effort researching this area, both from a domestic and an international perspective.

As a research site, this is one of the better ones we've seen. It includes pages and pages of analysis on some vital topics:

✔ The application of biometric technology. You can download a PDF book on this topic from the site, free of charge, at

`www.nationalbiometric.org/news_events_publications_std.php`

✔ International privacy laws with respect to U.S. applications of biometrics

✔ U.S. privacy laws with respect to the application of biometrics

✔ Published and emerging standards in biometrics

United States Department of Justice

The U.S. Department of Justice (DOJ) has a keen interest in biometrics and generally being able to identify and authenticate people — as well as a duty to the people to protect privacy and freedoms. To this end, the DOJ doesn't tend to take a stance on biometrics directly, but it does publish a lot of information about the laws of the land — and which are used to interpret how our privacy and freedoms are protected.

The DOJ Web site is a gigantic place to go looking for stuff, even if that stuff is sometimes nearly impossible to interpret simply. What we provide here are multiple links to privacy and biometrics-related links at the DOJ (and one from the department of defense):

- ✔ **National Institute of Justice**

  www.ojp.usdoj.gov/nij/topics/technology/biometrics/welcome.htm

- ✔ **Privacy Act of 1974**

  www.usdoj.gov/oip/privstat.htm

- ✔ **Office of Information and Privacy**

  www.usdoj.gov/oip/oip.html

- ✔ **Department of Defense Biometrics Task Force**

  www.biometrics.dod.mil

- ✔ **Privacy and Civil Liberties Office**

  www.usdoj.gov/pclo/

National Institute of Standards and Technology

The National Institute of Standards and Technology (NIST) is an enormously helpful resource for just about any information-security-related questions you may have. In Chapter 10, we discuss the NIST 800 series of documents; these cover everything from security controls and policy to specific encryption standards and their use.

As a nationally funded agency, the NIST's task is primarily to research and answer questions of this nature for other government agencies — but as a general rule, most of what NIST has to say is applicable to private industry as well, though you may have to trim off some of the government overkill. For example, if the NIST lists 17 separate security controls regarding authentication, your organization may need to use only seven or eight of them. Just divide by two and round up (just kidding — "use what applies" is the rule here).

Because the U.S. national government has a keen interest in biometrics and the applications of biometrics technologies, the NIST has devoted an entire section of its Web site to the topic at the Biometrics Resource Center Web site, located here:

```
www.itl.nist.gov/div893/biometrics/
```

This section of the NIST Web site rather broadly covers standards in biometrics including some of the interoperability and data interchange standards and testing. The site, however, also talks about test tools and applications that the NIST has developed as reference standards.

International Center for Disability Resources on the Internet

The International Center for Disability Resources on the Internet (ICDRI) focuses generally on the equalization of opportunities for persons with disabilities. We include its page on resources related to biometrics because we've noted an interesting contrast: The EFF raises concerns about denying equal access because of a lack of specific biometric traits — but the ICDRI sees biometrics as a potential *enabling technology* that will allow disabled people to interact with society on a more equal basis . . . *if applied correctly.*

The Web page is located here:

```
www.icdri.org/biometrics/biometrics.htm
```

It's really a launching point for links to other resources (so you get your money's worth in this Part of Tens) that describe how biometrics can be used to enhance a disabled person's access to society and in some cases how to avoid impairing that person's access when biometric systems are contemplated.

With all the currently available biometrics technologies, there will be a percentage of the population that will not be able to use the device due to a physical limitation. Blind people who may have perfectly good irises will struggle to use iris recognition because getting a good image requires that the eye be positioned by focusing on something for a moment. There are many other examples of this sort of thing, but these "exceptions" do not absolve people who are planning biometric projects from doing their best to plan around foreseeable issues. Resources that the ICDRI tracks on its site can help with that task.

findBIOMETRICS.com

This site is trying hard to be a complete resource for anyone preparing to undertake a biometric project. It has tons of information and links to information that will be useful in making decisions about your upcoming biometric project. For example, it does a great job of the following:

✔ Pulls together vendor lists categorized by biometric type (such as fingerprint or iris), or by biometric application (such as physical or logical access).

✔ Provides links to a large number of featured reports, articles, and news releases, all of which are good reading for someone contemplating a biometric project.

We really have just two caveats for readers who visit the site:

✔ **The site makes its money from advertising biometric products.** We are appreciative of the fact that sponsors as well as advertisements are clearly identified. As with any commercial interaction, however, users of the site should be aware that some of the biometrics technologies mentioned on these pages are helping the site pay the bills — and some are not.

✔ **Where biometric vendors are categorized by the application of their technology, HIPAA is listed as an application area.** The Health Insurance Portability and Accountability Act is most certainly a place where biometrics *can* be used to satisfy some of the requirements of Security Rule, we have two bones to pick with this link/categorization:

• **No biometric technology is "HIPAA-compliant."** That's a term without meaning in most cases. They don't use the term, but a link to technology categorized as "HIPAA" comes close to saying that.

• **Many of the technologies mentioned are there just because they provide authentication.** Authentication *is* one HIPAA security rule requirement. You might as well make that link just points to the "logical authentication" section and be done with it.

Caveats aside, we like the information and the site, and you can find them at

    www.findbiometrics.com/

The site's name should come as no surprise.

Third Factor Biometric Authentication News

This site is in a very simple blog form, and focuses entirely on biometric news. The articles are short and to the point, developed it would appear from press releases and other sources, but we like the simplicity of the site and its search capability (the unlabeled box at the upper right).

The site is owned and operated by AVISIAN, a company that makes its money primarily from consulting in the area of identity management, and uses their various publications in this topic area to promote the companies consulting practice. We have to say that promoting your business by providing a valuable information resource is a great way to get people to like you, and we appreciate their efforts.

Because it's bloglike and news-related, this site is a prime candidate for a Real Simple Syndication (RSS) feed. If you already use an RSS reader, the link you want is this:

```
www.thirdfactor.com/xml/rss20/feed.xml
```

If you want to read it with a Web browser, you'll want this link:

```
www.thirdfactor.com/
```

If you're not already using an RSS reader but read a lot of news online, you might be interested in reading up on RSS; it can be a real timesaver.

Biometrics Catalog

The Biometrics Catalog isn't really much like a catalog at all. According to its intro page it's free to use, U.S. government-sponsored, and a "database of public information about biometric technologies."

The site boasts some very handy features:

- A news section, which should be free of commercial bias (but with plenty of government bias, of course)
- A comprehensive searchable list with links to U.S. government publications about biometrics

✔ A gigantic database of research reports, totaling more than 6,000 documents, also searchable

✔ A section with links to papers and reports on biometrics privacy

✔ Conference presentations, legislative reports, and lists of commercial products and vendors

This site, combined with the lack of commercial bias, is an excellent searchable resource for folks wanting to understand a bit more about what's happening in biometrics. It's not as focused as some others are on providing answers to questions such as, "What should I buy?" It will, however, help educate you to a point at which you can answer that question for yourself. The "catalog" can be found here:

```
www.biometricscatalog.org/
```

John Daugman

We thought a bit about what to put into the last slot here in the Web sites, and settled on John Daugman's Web site at the University of Cambridge. Unlike most of the other biometrics, which owe their origins to many sources and parallel development, iris recognition — provably one of the most accurate biometrics available — owes much of its history to just one man: John Daugman.

Since all commercial applications of iris recognition use algorithms originally developed by John, and iris recognition is poised to play such an important role in many of the high-volume biometrics installations that value ease of use, throughput, and accuracy, it seemed fitting to include his Web page as an important biometrics Web site.

John's site is updated with current work in computer vision, iris recognition, and good data from some of the more interesting installations using his algorithms. You can find him at

```
www.cl.cam.ac.uk/~jgd1000/
```

Chapter 14

Ten Essentials for Biometrics Success

*T*here are a lot of ways to measure success, and biometric projects pose some interesting challenges that require a careful understanding of what it means to "succeed" — and exactly how we will measure that. If we're too narrow in our definition, we risk being spectacularly unsuccessful in some aspect of the project — even while still meeting our general success criteria. Accidentally locking everyone out of the building by setting the False Accept Rate (FAR) too low would be a good example of this. If your success criterion was to keep unauthorized people out of the building, you clearly met it; if everyone else is also left standing outside in the rain (while you hide in the server room relaxing the FAR a bit), the project might not enhance your bonus this year.

Align with the Goals of the Organization

No project can succeed if it doesn't accomplish something meaningful for the organization. In most fields, that's never a problem, because nobody ever even proposes a project without very specific organizational goals that the project will accomplish. IT, however, sometimes seems to see a need for new technology when there are no clear tactical or strategic benefits to the organization beyond "It will do X better and faster than we're doing it today" — especially if what you do today is good and fast enough.

Don't get us wrong. Both authors have worked in IT, IS, MIS, DP, or whatever name you want to put on the management of information technology for quite a long time; we know and understand the need to keep current and not let information technology in the organization go stale. We also know that being on the bleeding edge means that someone or something is going to bleed — usually the very business processes that support our paychecks. There is a time and a place for using the coolest, newest version of some cool new technology — but usually that isn't in production systems with a low tolerance for downtime, odd behavior, and the whole list of symptoms from a television drug commercial.

Biometric projects should be trying to solve an *identified problem* in an organization *for which biometric technology is best suited*. Such identified problems crop up when organizational needs go unmet in (for example) these categories:

- ✔ Identification or authorization that is tied directly to an individual and not sharable
- ✔ Strong authentication that doesn't require memorizing long, difficult passwords
- ✔ Identification or authorization that does not depend on any level of trust with the subject

Above all, successful biometric installations are usually accomplished as a response to an expressed need from *outside* the information-technology organization. The expressed need may not mention biometrics, but it will name requirements that biometrics can best fulfill.

Consider and Address Privacy Concerns

One measure of success for a new technology is that people using it do not resent its being forced upon them and feel that it's affecting their lives in a negative fashion. Unfortunately, if you're not careful to consider the potential privacy concerns associated with a new biometrics project, that is exactly how many users will feel about it.

Are their concerns valid? Is their privacy really being invaded and compromised by the system? Being able to answer "no" to each of these questions doesn't really buy you anything within your user community unless you can also explain why and how their privacy is protected — in terms that they understand.

For example, telling a sales manager that the fingerprint scanner is only stor-ing a hash of specific minutiae from the fingerprint (and not a photographic image of the print) may not mean much to the sales manager. Explaining that it's not storing an image at all, but just some numbers that could never be used to re-create the original print, might make more sense. Further explain-ing that it's one-way-encrypted (like boiling an egg — no way to unboil it) and the only thing that the database is good for is comparing a new scan (also boiled) to the previously boiled version of an old scan might give them a handier picture of what's happening.

For the following two reasons, we recommend that you never *ever* dismiss users' concerns as "not based on the facts." Here's why:

- **Reality or no reality, the users are concerned.** Remember that a user's perception is always right, even if it's not based on facts. Telling them they don't know what they're talking about will not alleviate their con-cerns — not even a little bit.

- **It's your job to make sure users have all the facts before them, and in a form they can understand.** If they don't, it's not their problem, it's yours. In some ways, it's like editorial feedback to an author. Even if what you wrote is perfectly correct, the fact that the editor didn't under-stand is a big indicator that you need to try and explain it another way.

Survey the Users

Although privacy may be the most common concern for users, it's by no means the only one. Since your success is based at least partially on their acceptance of the system and comfort in using it, we suggest that you have some discussions with representative users about your candidate solutions — and about the impact of a biometric installation on their work.

As with privacy, it's important that you take their concerns seriously; don't just dismiss ideas that are clearly uninformed. We're reminded of people who complained in the late 1980s that computers and cellular phones were going to eliminate direct human contact and impose great barriers to real comm-unication. Way back then, we saw the great potential for better electronic communications having the opposite effect — bringing together people with similar interests and allowing diverse groups to communicate who would otherwise have ignored each other. Even though we didn't agree with their assessment, we worked with these users to help them see positive uses for the tools provided. For all we know, some of those users went on to create some of the first social-networking sites . . . nahhhh.

As we mention in Chapters 9 and 10, there are lots of things to think about in a biometric installation that have nothing to do with protecting assets with biometrics — and have everything to do with people interacting with something new. For example, touching a fingerprint or palm-vein reader imparts exactly the same amount of germs as touching a doorknob on the same associated door; because the biometric device is new, however, people think about the implications of touching it right after the sneezy guy in front of them — and want to know *what you're going to do about it.*

If you perform a wide survey of your potential users, the one thing we can almost guarantee is that you'll see questions crop up that you never expected to hear.

Stick to the Plan

Part III is where you get a plan to stick to; it's all about putting a formal process together to select a solution, plan its implementation, put the solution in place, and use it for its intended purpose. If you're a little rusty on Part III, skip back there for a refresher. We'll wait.

Back already? It's good material, maybe you'd like to take some time and go over it all again? All kidding aside, some of the more spectacular failures of any IT implementation come not so much from a poor plan as from failure to follow an excellent plan. Both of us tend to get called into projects when things are not proceeding well; we've each been around the block a few times, and have reputations for being able to fix projects on the edge of failure (we fondly call these "rescue jobs"). When we're called in on cases like this, we see the same scenario about seven or eight times out of ten: The project manager hands us a great plan that details the particulars of the selection process, testing, and implementation — a plan that was abandoned a few months earlier because "the landscape changed." It's as though they've thrown their oars into the water and then wonder why they're drifting off course.

About one time in a hundred, the basic parameters for the original project actually *did* change, for good business reasons, partway through the execution of the original plan. In those cases, you really need to *completely review the whole project* to make sure it will satisfy organizational objectives. The other ninety-nine times out of a hundred, what really changed was more like staffing for the project, or timelines, or (our favorites) funding and priority. In those cases, the original plan is still the path to success; you'll need to make sure that management knows that new obstacles have been placed in the way of your project. They should also be made aware that the original plan was carefully considered, and the resource changes that happened mid-course are threatening the eventual success of the biometric project (which is still directly linked to the success of specific business objectives).

Be Flexible

Be flexible is really the corollary to "Stick to the Plan" and not the opposite. A really great plan will anticipate some of the kinds of kinks that come up in the execution of the plan, and allow flexibility without deviation from the final project goals.

A lack of flexibility in the plan, or in the execution of the plan, can be almost as damaging to the success of a project as a total lack of a plan. In a perfect world, the circumstances of your biometrics project will exactly match the assumptions made during the planning process, and every detail of the plan will execute exactly the way it was committed to paper — without deviation, delay, or lack of crossed *t* or dotted *i*. We've never lived in a perfect world, though; when implementing a technology that's as new on the scene (and as fast-developing) as biometrics, the chances of everything happening *exactly* the way we expect are near zero.

How do you balance adherence to a good plan with flexibility? As a part of your plan, implement the following concepts:

- ✔ A formal change-control policy with a review process for proposed changes to the plan
- ✔ An approval process for stuff that makes it through review
- ✔ A process for examining the effects of the change once it's in place

If your plan anticipates and accommodates changes in a formal way, changes can be a part of the plan.

Research the Problem

Here's an effective way to build on the idea that the whole reason you're contemplating a new biometric solution is to meet an organizational requirement: Express the problem in familiar business terms, like this:

We need a way to accurately and effectively identify personnel in access-management situations.

There is an excellent chance that your organization is not the first one to encounter such a problem, and that others have already done a fair amount of work toward understanding the nature of the particular problem you're trying to solve. And they probably have something to say about it.

In some rare cases, the problem may be entirely unique to your organization and nobody has anything to say about it. You can still increase your chances of success greatly if you do some research into the basics of the problem before you start thumbing through the catalog for solutions.

For example, suppose you work in health care and you've been told that the doctors are sharing their credentials with their administrative staff, so you never know who really logged in to look at a patient record — it always shows up *as* the doctor. (Maybe this situation came to a head when the systems showed that the doctor accessed a patient's information when you know the doctor was on a cold-air balloon flight across the Pacific Ocean.) In this case, you know that biometrics is a good way to tie authentication to a particular individual, so you've started thinking about some kind of biometric solution to this problem.

Additionally, suppose you like fingerprint readers because they're inexpensive and readily available; you've tossed out that idea, though, because many workers have to be authenticated while wearing latex gloves — and clearly that won't work. A little bit of online research will show that there are actually several fingerprint readers that work just fine through latex gloves — because *many* people have found themselves in your situation, and manufacturers have moved to fill the need.

In some cases, you will also find while researching the basic problem that other people have solved the same problem without the use of biometrics. Depending on the solution, that can be the best discovery of all, since it might save the organization the expense of the biometrics project entirely. Sometimes success in information technology means finding another, simpler way to accomplish the original goals that may not even be technology-related.

Research the Solution

Given that you have researched the problem and that you understand the parameters of the problem well, the next obvious way to improve your chances for success is to identify some potential solutions and do some research into each of them to really understand the differences and potential implications of choosing each of them.

For example, you know you'd like to use a technology that has as little impact to the user as possible, but is accurate to 99.99 percent, so only one person in 10,000 would be accepted when he or she shouldn't have been. You figure that since you're using this solution for authentication rather than identification, chances for a false acceptance are even lower than the 1-in-10,000 and that will work for you. While identifying several nonintrusive methods, you decide you like facial recognition, gait recognition, and iris recognition as your favorite candidates. You toss out gait immediately, because it doesn't have the accuracy you require and is typically used for identification rather than authentication. After further research, you also toss out facial recognition — even though the manufacturer "guarantees" the level of accuracy that you require — because your research led you to discover that there's enough similarity within some family groups to skew the False Acceptance Rates for facial recognition, and the installation is a high-tech home with specific access rules for family members. Iris recognition for the game room it is!

A complete understanding of the peculiarities of the biometric solutions you're considering is a requirement for success — and biometrics sometimes have some pretty odd quirks.

Don't Get Fancy

Given two solutions, each of which solves the same problem equally well, which solution is the *least* likely to get you paged at home on a weekend in the middle of something fun? The answer is always the *simplest* solution.

Biometric solutions can be as complicated as a multi-site, multi-modal system that does identification, tracking, and authentication — watching your gait to guess who you are while you walk up to the iris-recognition system that performs a faster search of the iris database. (What do you *do* for a living? No, on second thought, don't tell us.) They can also be as simple as a finger swipe on a laptop to log you in.

When purchasing technology, there's always a tradeoff between two competing concerns:

- ✔ Getting something sophisticated enough to meet the needs of the organization, even as its anticipated growth sprouts additional needs
- ✔ Providing a simple, inexpensive solution that meets the *current* need

Both are valid viewpoints, but you should ask yourself the following questions:

- ✔ Will you outgrow the technology before it becomes technologically obsolete anyway? If not, buy the cheaper, gets-it-done-now version.
- ✔ How hard and/or expensive is it to upgrade to something better if and when you need to?

✔ How good is your organization at predicting growth and future needs? Are you confident you know how long the basic system would last?

✔ What are the chances that the next iteration of this problem will require a different solution entirely — in which case, you'll have wasted the money spent on flexibility and growth planning?

The simpler you can make your biometric solution — while still meeting current needs, anticipated growth, and additional requirements — the better you'll sleep at night, and the more money will be available for bonuses at the end of the year.

Talk to People with Working Installations

You might think of this as an extension of the "Research the Solution" topic earlier in the chapter, but it's actually much more than that. Unless you're riding the bleeding edge and implementing the newest technology well before anyone else, there's an excellent chance that you can get in touch with other people who have already made huge mistakes in implementing biometric systems — and find out what those mistakes were so you can avoid repeating them.

Of course, you should also include a discussion of exactly what those folks did right, and try to re-create those moments if they apply to your project. Be aware, though, that most people don't remember when everything went perfectly nearly so well as they remember when parts were flying through the air and the fire trucks were still ten minutes out.

The problem with starting up a conversation about someone else's painful moments is that if they don't already know you well, they might be a little reluctant to recount they day they almost got fired. The best way to resolve this dilemma is to get to know a bunch of people who do jobs similar to yours, long before you need to ask them these questions. There are lots of groups out there organized around meeting and discussing information security among peers. We strongly advise that you find one you're comfortable with, and spend some time getting to know those folks; not only will they become a valuable information resource, you can rest assured that somebody somewhere *actually knows what you're talking about.*

Do Not "Fire and Forget"

The final item in this list of ways to succeed is to pay attention to how your biometric system is doing *after* the installation. For most projects, you'll spend a lot of time preparing for the first day of operation, making sure that it will come off without a hitch, making sure that all the work you've done in selecting and implementing your biometric solution comes together in a blinding flash of absolute perfection. (Good luck with that.)

Keep in mind that for most of the users of the system, the first day is just that — a day like any other day, but one that required them to learn a new biometric system and start using it. Depending on the system, they'll go on to use it several more times that day, and then hundreds of times over the next few months. The fact that it worked perfectly on the day it was installed will pale in comparison to the multiple annoying failures they have to endure after that brief shining moment — unless you're paying attention to the system right along.

Not every technology will immediately start to deteriorate while you watch, but a lot of them do. A successful biometric installation will be one that catches new problems quickly as they crop up, and deals with them efficiently. In most organizations, once you design and build a solution, you own all the problems attached to that solution until you move on to new employment — and in some cases, users will still track you down and ask questions.

 The perceived success of the system in the minds of users will be how well it works over time, so do yourself a favor and heed these two tips after that first day of deployment:

- ✔ Watch the installation carefully for the first month or so, noting trends in false rejects, failures to enroll, and any other biometrically unique attribute that seems to be drifting or careening out of control.

- ✔ Watch resource utilization at the levels of individual computer *and* network; make sure they're running well and won't exhaust their available resources any time soon.

Chapter 15

Ten Biometrics Scenarios to Plan For

In This Chapter

▶ Surveying some common uses

▶ Watching out for pitfalls

▶ Dreaming up uncommon uses

*B*iometrics encompasses a pretty large field of study, and there are a lot more use scenarios than you might think. In this chapter, we consider some of the common use scenarios, some possible pitfall scenarios for biometrics, and some uncommon, but interesting, uses of biometrics technology.

Alternate Entryway Access

If your building has a lobby with either a security guard or receptionist, you may have no need for a biometric entry control for the lobby. Depending on the size of the company and the size of the building, the person at the desk may recognize everyone who has a legitimate need for access. Even if that isn't the case, there isn't really any way for an unauthorized person to know for sure which personnel the person at the desk knows or doesn't know — so a human being is still a pretty decent deterrent to attempted unauthorized entry. Only in the most risk-averse companies with very high security requirements will you see biometric entry systems in the lobby or front entrance.

Side entrances and parking lots are a different matter entirely. Most side entrances and parking-lot elevators have no human supervision, so anyone who happens to know the door code from watching someone else punch it in (or has a proximity card picked up from the parking lot) would be able to gain access to the facility.

Keypad insecurity

Keypads can be reasonably good access security in some circumstances, but there are a lot of things that can go wrong with keypads to make them far less secure. For example, if everyone uses the same code, not only is it likely to get scribbled on a scrap of paper and lost, the actual keys used for that shared code will become more worn unless changed weekly. Our favorite keypads to break into are the shiny-surfaced ones or touchpad screens, since you can see what keys folks are using by just adjusting the lighting. You may not know the order, but you've dropped the number of combinations you have to try to a manageable number.

 If you use even a simple fingerprint scanner at side doors or parking-lot elevators, you greatly reduce the likelihood that someone can get into the facility with stolen credentials.

High-Security Hosting

In some cases, even with a security guard at the door, so many people require access to a facility — and the access list has such high turnover — that it's impossible for even for a great security guard to remember everyone who should be allowed in. Even someone with a great memory for such things can't be expected to know (say) that Johnny was just fired an hour ago, and we don't want him allowed back into the building, especially if he's mad.

There are probably many examples of this sort of thing, but the one we encounter most often is entry access to computer-network hosting facilities. In these scenarios, a hosting company offers rack space, network connections, environmental controls, and power to customers who want to host services at such a facility instead of building it themselves. Since the hosting provider has a lot of customers, and one of the things they are providing to customers is physical security for the servers, they carefully control who has access to the hosting floor.

Each customer submits a list of people who are allowed to enter the facility on their behalf, updating that list as people are hired and as they leave the company. The hosting company, in turn, is expected to only allow authorized persons into the facility. The hosting company has really only two options for accomplishing this goal:

✔ Keep a list of authorized persons and manually check each incoming person's identification (usually a photo ID).

✔ Keep the same list in the form of *enrolled biometrics* and have the system sort people out.

Either way, you're using biometrics, but manual verification takes too much time (besides, it's not as impressive when you're showing the facility to prospective clients).

We've used the example of a hosting company here, but really any place that performs high-throughput authentication for third parties is a good candidate for this scenario. If you're in that situation, just make sure that the processes for adding and deleting authorized personnel are simple, efficient, and prompt.

Port-of-Entry Identification

Port-of-entry scenarios are a bit like high-security hosting, only on a massively larger scale, and without the advantage of a master list of authorized persons. The traditional port-of-entry system involves the use of a hard to duplicate identification (a passport) which users carry with them. Manual biometrics are accomplished when the officer at the gate compares the passport picture with the person presenting it. This system depends to a large degree on how hard it is to modify or create a passport without being detected.

One of the simplest ways to use biometrics to assist in this process is to include electronics with the passport itself so it has these two features:

✔ An electronic key to verify that the information on the passport was encrypted using a key belonging to the country of origin.

✔ Encrypted biometric data that can be compared to the biometrics of the person presenting the passport.

This system isn't perfect, but it does make the data somewhat harder to forge — and still doesn't require a master database.

Another port-of-entry concern is the identification of persons known to be on watchlists, or those who have been expelled from the country for some reason. In these cases, it's often possible to collect biometric information from the subjects — with or without their cooperation — and then use that information to identify them as they try to cross the border. This scenario does require a master list of persons and their biometrics, but it's much smaller than the list of *everyone on Earth who might decide to travel*.

Law Enforcement

We don't always think of law enforcement when we consider biometrics scenarios, but law-enforcement uses probably constitute 90 percent or more of the known applications for biometric technology. Law-enforcement agencies use biometrics in the same way as any organization with a need for high levels of security and accountability, but their use of biometrics for authentication are much the same as those in use by everyone else.

It's the use of biometrics for *identification* that distinguishes law enforcement from most other use cases. For many years now, law enforcement has used fingerprints collected at crime scenes to later establish the identity of the person holding the knife, gun, or pogo stick when the crime occurred. More recently, police have been able to use blood and other biological evidence they collect to identify persons through DNA comparisons. Although these uses are also part of biometrics, they're pretty well known by anyone who watches crime shows on television.

The more interesting biometric scenarios in law enforcement these days involve the use of images collected from sources such as ATM cameras, surveillance video, and private citizens with cameras to identify people using facial recognition, gait recognition, and even body-mass-and-movement biometrics. Although many of these techniques are not nearly accurate enough for an identification that would hold up in court, they are useful tools to help eliminate some suspects and spotlight others to guide an investigation.

If you ever wonder just how much potential biometric information is being collected from you on a given day that could be used in this way, take a look around when you're walking through parking lots or checking out at the store — and smile nicely to the ATM camera behind the dark plastic. Some cities in some neighborhoods even have police video cameras watching public places.

That Biometric Doesn't Work for Me

Depending on the size of your user population and how long you operate a biometric solution, this scenario will present itself sooner or later. Some of these situations you probably anticipated — where, for some reason, the specific biometric you're capturing just doesn't apply to someone because the right hand (or other required body part) is either missing or just doesn't conform to the parameters that the biometric system expects.

The scenarios that will surprise you (or *would* have if you'd skipped this chapter) are where everything physically seems okay, but the biometric measurement nearly always fails for a specific person. The problem is that there's a pretty broad variation across our species, and some measurements make a lot of assumptions that are based on statistical norms. Two examples of particular biometrics methods that don't work well for all people are

- **Fingerprinting:** Some people just don't fingerprint well. A good print is taken from a finger with deep furrows and good color contrast — but if your furrows are shallow or your fingertip color isn't contrasting well enough with the light source, it could be almost impossible to get a usable fingerprint from you.

- **Signature recognition:** This is another example of a biometric that works well for some people and not for others. Both of your authors have a constant battle with keeping their signatures consistent enough so banks and credit-card companies don't start rejecting their transactions.

About all we can say about this scenario is that if you're in charge of making the biometric solution work, you will need to make accommodations for people who aren't able to use the system.

I Object to the Invasion of My Privacy

We don't have any silver bullet for this objection to the implementation of biometrics, but it's a scenario you need to be prepared for. Keep in mind that not everyone will have the same perspective as you on biometrics, and there are a lot of well respected security analysts that have gone on record objecting to the widespread use of biometric information. People who express this concern are not crazy Luddites who want the whole *computer society* thing to just go away (well, okay, not *all* of them are). Some are thoughtful people who have put a lot of that thought into the subject — and feel that collecting and storing biometric information about them is a personal violation of their privacy.

In such cases, it may help to get to the practical root of the problem. Ask folks exactly what bothers them about the fingerprint scanner and database; the idea is to fully understand their concerns and try to address them. Here's an example of the sort of worry you may hear voiced: What if the stored biometric data could be used to create a false version of the same biometric data (such as fingerprints) — and somebody does that, goes down to the dealership, and uses that data to okay the purchase of a new Ferrari? If your company and the Ferrari dealership use the exact same technology, and the car thief can steal the fingerprint database — and *then* inject the correct record into the transaction while the dealer watches — then the thief *might* drive off the lot with your user's retirement.

Scary? Sure. But the actual risk of that happening is small, for the following reasons:

- ✔ **How biometric data is stored.** Often very specific to the equipment used, making it impossible to recreate a print or other biometric original, just a record of the information details.

- ✔ You have to inject the stolen biometrics into *exactly* the right place in the system at exactly the right time to fool it into thinking you're someone else.

- ✔ You still have to convince the dealer, the credit-card company, and a lot of other folks that you're someone else by using a fake ID and some fancy talking.

It's not impossible for the worst case scenarios to come about, but for now there are far easier ways to steal someone's identity than stealing and using biometric information — and, for that matter, easier ways to steal biometric information than hacking into the database (such as photographing someone or picking up fingerprints from a discarded drink container). We discuss addressing users' concerns in this area in greater detail in Chapter 8.

If the argument is more philosophical than practical, we can't help you much there except to suggest debate classes.

We Found Malware on the Biometric Database Server

There's a raging debate in information-security circles as to exactly what it means to discover malware on a server that contains sensitive information. This scenario happens hundreds or thousands of times each day around the world, and it may trash you least if you've already considered what it means to you — and how you would prefer to handle it.

The problem in this scenario is what you *don't* know: Suppose you've just discovered that a program with some sort of ill intent has been installed on a system containing your biometric database, and you don't know for sure what that program code may have done, or had access to. In the worst-case scenario, it granted hackers direct access to that server (and possibly the rest of your network) and the hackers downloaded all the information to see what they could do with it. All the information on that system may be in the hands of someone that intends to do you harm.

The debate mainly circles around the concept that the existence of malware on the host does not imply or in any way guarantee that information was compromised. If the information was not compromised, all you need to do is eliminate the malware and you're good to go. The other side of that coin is that since it *could* all be exposed, you need to act as though it actually was.

Our perspective on this scenario is that it all depends. Ask yourself these questions:

- ✔ What are the security requirements for the installation?
- ✔ What is the legal, regulatory, and public relations environment for the organization?
- ✔ If the malware is well known to the antivirus/antispyware community, do you also know what it really does?

If you can answer those questions, you'll have a much better picture of what to do in this scenario.

Biometric Readers in Objects

Since biometrics can be a good way to grant authorized access to stuff, why not put biometrics into actual stuff we want to control access to — such as, say, credit cards? It might sound a little futuristic and farfetched, but a Danish firm — Scanecotech A/S — developed exactly this technology. More interesting than its specific application though is the thought that if you can embed biometric readers into something as small and thin as a credit card, it's possible to consider using biometric authentication for almost any object or device.

For example, some of the high-end portable music and video players are magnets for theft because they are small and very valuable, and the same goes for portable GPS receivers, cell phones, and cameras. It would likely deter theft a bit if the device wasn't accessible without a biometric login, and bypass techniques all set off the autodestruct (nothing big, just a good memory meltdown). If a thief had no chance to get a working unit, there's really no reason to steal it.

In the case of cell phones, there's an added dimension to the idea of having a biometric reader on the phone, which is the ability to authenticate transactions enabled by the phone either as a network connected device or as a cellular phone.

As a connected network device, the phone can operate just like any other biometric scanner you might have on your network, only this one is portable. A transaction that was initiated via a network-aware program on the phone or a web browser on the phone could be authenticated by asking the user to present their biometric information at the proper time.

As a cellular phone, a typical phone tree transaction ('press 2 to buy the Ferrari') could at some point also be authenticated by asking the user to present their biometric information. In the case of a cellular telephony transaction, the biometric authentication would likely trigger a DTMF number sequence (Dual-tone multi-frequency — the sounds the phone makes when you touch a number, which would work like touching the numbers on the keypad) that corresponds to an authorization code.

Behavioral Biometric Driving Sensors

Some kinds of behavioral biometrics can identify someone through that person's interaction with a system entirely unrelated to identification or authentication. Examples include driving, typing, or even just watching how you move a mouse to accomplish routine computer tasks. Depending on the measurement, the sample size, and the reliability of the specific biometric, these identification events can be more or less accurate.

Sometimes accuracy is not critical to a particular use of biometrics, and the more important concern is to prevent the biometric measurement from having any significant impact on the user. As a good example of that, imagine a biometric system built into a family car. The system watches how you drive, brake, signal, and even steer so it knows and understands who is behind the wheel. Since the list of enrolled drivers is small, the system doesn't have to make very many comparisons, and if the system's purpose is to help decide what music to play from the on-board mp3 collection, making the wrong decision won't have all that big an impact (unless it decides that you "are" your teenage son and chooses music and volume levels to match . . .).

In the driving scenario, you could even have the system perform more complex decisions such as ensuring that there is someone of adult size in the passenger seat when the teenager is driving, or having the inboard cellular phone start calling you when someone completely unknown seems to be driving the car. Doubtless we'll see other scenarios where biometrics can be applied for identification purposes and then used to perform low-impact but helpful actions.

Biometric Neighborhood Watch

Have you ever lived in a neighborhood with a well-organized neighborhood watch? Some people like the idea, and it makes them feel more secure knowing that neighbors are keeping an eye on things even when they're not at home. On the other hand, the very idea that folks are watching what's going on and taking notes really creeps some people out.

We haven't seriously seen anyone considering the idea of a biometrically enhanced neighborhood-watch program, but as a concept it actually might help alleviate the creepiness factor a bit. Instead of retired folks with binoculars sweeping the 'hood for strange activity, you could have video cameras watching public places and identifying strangers. A learning system would even figure out that people who walk through your neighborhood to the bus stop every morning at 8:00 a.m. are not strangers, but those same people walking down the alley at 2:00 a.m. are strange *to that place and time.*

The monitoring and collection of this kind of data is one thing, but it's what you do with it that makes this a scary neighborhood with cameras watching your every move, or a safe place where criminals learn not to hang out. For example, you could choose not to even record video that depicts well-known people in places where we expect to see them — and only record strangers or unfamiliar people in particular places, at times when you don't expect to see them. Further, the policy could be that even the footage of strangers doesn't get looked at until someone notifies the watch that something specific happened and there is footage recorded for that time and place.

Chapter 16

Ten Benefits of Biometrics

. . . .

In This Chapter

▶ Listing what we know

▶ Dealing with the "Un"s

▶ Evaluating business benefits

▶ Getting mesmerized by coolness

. . . .

*A*ll too common when a new technology becomes popular, it gets plugged into any technological nook or cranny that anyone with a budget can imagine. In most cases, that frenzy leads to some level of overload — for information-technology professionals and for users (is anyone else tired of using Flash to fill out a simple Web form yet?). Biometric technologies are not immune to over-application, but there are some clear benefits to using biometrics in the appropriate situations.

Here, we list benefits to authentication and identification that are unique to biometrics, as well as benefits with wider impact than just identification and authentication.

Cooperation Not Required

In some cases, you need to identify or authenticate persons who may not be much interested in helping you with the process. For example, you might be screening for wanted felons at the entry to a ballgame with facial recognition, or looking for known assassins at a government function using gait recognition from surveillance video. The key here is that you can accomplish the identification while the person being identified is actively trying to conceal it.

This is a characteristic that's really unique to biometrics in that biometrics are something that is a part of you, and about the only way to escape biometric identification is to somehow change that part of your appearance or actions. Some biometric techniques — such as facial thermography and certain behavioral biometrics — are pretty good at seeing right through disguises. That's because they aren't looking at surface features at all.

In the "not as beneficial to society" column is the idea of using this same concept in public spaces (such as shopping malls and retail stores) to identify previous customers and send a salesperson over that way to do some directed marketing. Imagine the following conversation from a salesperson five minutes after you walk in the door of a nationwide clothier, 2,000 miles from your original purchase; "How's that new black suit you bought last February, Mr. Smith? It would really look great with these silver cufflinks, don't you think?" Creepy, to say the least.

Guarantees Physical Location

Biometric technology requires the presence of the actual person associated with the specific biometric measurements at the biometric sensor. As long as you know for sure where your sensors are, you will always know exactly where someone was when that person was authenticated (or identified himself or herself) using that sensor. Because of that requirement, it's possible to use biometrics in places where you have to make sure the users cannot later claim they "were never there" and "did not authorize" actions taken in their names. This concept of *non-repudiation* is important to financial transactions, military orders, and really any place where someone might suffer a change of mind later and not really *want* to accept responsibility for a damaging action that *someone* (guess who?) performed.

Biometrics can accomplish the task of non-repudiation in a couple of different ways — each creating a different level of confidence in the action:

✔ **Using biometrics as an authentication method to prove that the people logging in are who they say they are.**

After logging in, users then go about their tasks, and the system makes three assumptions:

- The person who logged in is the authenticated user.

- The person who logged in is present.

- The person who logged in is responsible for each action taken by the authenticated user.

This method has a weakness: The authenticated user may have walked away from his or her workstation after the initial authentication, allowing someone else to take action using the authenticated identity. For high-value and high-importance transactions (like stock trades), you might have the user authenticate *again* to approve the transaction.

✔ **Special actions that require non-repudiation require the user to submit biometric information when initiating the action.**

This method has a better confidence level because notwithstanding the initial login, ensuring that the expected person is still in control of the workstation at the time of the special action. It's hard to claim (for example) that you "weren't there" when 30,000 shares of stock were purchased if we can prove that your living finger, eyeball, or hand veins *were* there.

High-Throughput

When you consider the many alternative methods used to authenticate or identify people, especially where there is some danger of fraud or false identification, biometrics quickly rises to the top of the list for being able to process lots of persons' identities very accurately.

Passwords are not typically used in such scenarios, because they can be stolen or shared. Also, typing in a password of sufficient length or complexity takes a lot longer than swiping a finger or palm against a scanner — and that's not even taking into account *shy password syndrome*, where people flub their password two or three times in a row, just because they're under pressure or being watched.

Manual comparison of photo ID is the only other method that is typically used in high-throughput scenarios where accuracy in authenticity is important. First, even manual comparison to a photo ID *is* a form of biometric authentication — it's just not quite as accurate or as good at spotting deception as a good fingerprint or iris-recognition system.

High-throughout *identification* systems are another story entirely. When the identity of a subject is assumed to be unknown and the population to be searched is larger than a few dozen people, automated biometric identification systems are really the only choice that works. For instance, spotting known terrorists via facial recognition at airports and other transportation centers — or making sure that everyone walking around in a secure area is really authorized to be there — is possible with gait, facial, or even iris identification.

Unforgettable

Most people have more than one account with corresponding usernames, passwords or Personal Identification Numbers (PINs). At the very least, if you aren't retired, you have a work login and a PIN for your ATM card. For people who work in information technology, the number of accounts and passwords to remember can be closer to seven or eight, with no real upper limit. If you routinely shop online or participate in members-only online communities, you could have dozens of accounts.

The problem with having lots of accounts is that there's a strong temptation to do something that most security folks would consider to be very insecure — namely, to use the same password on as many of the accounts as possible. The problem with this practice is that if attackers can figure out the password for *any one* of your accounts, then they automatically have access to many (or all) of your accounts.

The best practice is to choose complex passwords for each account that aren't related to any human language (nope, not even to Orc, Klingon, or Hobbit), *never write them down anywhere,* and *change them periodically.*

A real benefit of using biometrics to authenticate users instead of passwords is that the user gets to use the same "password" for multiple accounts without compromising security. Since the biometric is always with the user and really unusable by anyone else, an attacker doesn't gain anything by knowing that the same right index finger, eyeball, or hand-vein pattern is used to log in to multiple accounts. Sure, an *injection attack* that steals the data as it's being transmitted from the sensor and then re-injects the data into another system might be successful, but if your attacker is that sophisticated and has that much access to your network already, it's safe to assume you're already using encrypted connections for the biometric scanners, right?

The use of biometrics allows users to have multiple accounts and use the same biometric measurement for each, or even different biometrics for each and still not have to remember multiple passwords. Making it easy for users to operate in the most secure fashion is always a benefit to both the users and the organization.

Unlosable

Not all forms of authentication are passwords. Some are physical, such as

- ✔ Proximity cards that activate door entry systems when waved over the sensor pad
- ✔ One-time password tokens that display a password that can be used only once
- ✔ USB keys that allow access to systems when they're physically plugged in

Unlike these objects — which might accidentally end up in the washing machine, dog's chewy-bits pile, or at the bottom of the ocean, lake, or coffee cup — your biometric measurements are much harder to lose. Another plus: They're pretty much never just misplaced (of course, maybe the phrase "he'd forget his index finger if it wasn't screwed on" will catch on in the age of biometrics).

The various situations mentioned in the preceding paragraph have all happened to the authors, so we can say from experience that losing the item that authenticates you to various aspects of your job is inconvenient at best and semi- to completely disastrous at worst. For example, while watching the USB key do a triple flip from the center console of the car into the coffee (intended to keep him awake at 2:00 a.m. while he fixed the Web server for a client), Mike was reminded that a good *biometric* server-access system would have meant *not* having to call his client at 2:00 a.m. on a Monday.

Unsharable

People share access codes to systems. You can explain until you're blue in the face that it's not really a good idea and that your organization requires individual accountability for each person and the responses will range from: "Yeah, but it's more convenient to give my admin my password than to figure out how to give him access to the files he needs . . ." to "I'm busy trying to make money for this company, and this person (with whom I share my account) is helping me. How important can this really be?"

In some environments — health care, for example — sharing accounts and access codes is *legally forbidden* and punishable by fines (or even stronger sanctions if the circumstances warrant). In any organization, at the very least, it's a sloppy practice that can lead to serious confusion when trying to track the actions of an individual for personnel, forensic, or other reasons. The mechanisms for sharing are many, including telling others your password or letting them borrow your ID badge or authentication token.

One of the few ways to effectively combat this practice technologically is to use biometrics as a primary authentication method. Sure, you can still allow someone to use your login by walking over and logging them in, but at that point it probably becomes more inconvenient to share a login than to just share the required resources the right way.

Cost Reduction

Biometrics systems are not simple or inexpensive to install, but applied correctly they can realize significant cost reductions in several areas. Two places where you can look for these savings are

- ✔ **Improved compliance to company policies:** Companies that pay hourly frequently have a problem known as *buddy punching*. This is the practice of having one of your buddies grab your timecard and punch you in, even though you're running a little late for work. Both of you punch in promptly at 8:00 a.m., and you arrive at a more leisurely 8:55 to begin your shift. In the mean time, the company is paying for an hour of your time that you were not really at work. This is a common enough problem that biometric timecard systems are a readily available item. Employees might grumble a bit about trust, but it's hard to argue against the point that these systems are completely fair to both sides.

- ✔ **Reduced workloads for some IT areas:** Another way that a company can see cost savings by using biometric systems for authentication is through a reduction in time spent on password resets or replacing one-time password tokens. Every time a user forgets his or her password (or flubs it enough times to get locked out of the system), a phone call to the helpdesk is involved. The helpdesk then has to verify that callers are who they say they are, and then spend a few moments resetting that pesky password or account. In the case of a lost token, there's a direct cost of around $70 to $120, and a somewhat longer, more involved call to the helpdesk.

By using biometrics, you eliminate the forgotten-password calls and lost tokens — and should greatly reduce the number of lockouts (or you need to adjust the system for too many false rejections). Fewer calls to the helpdesk means less helpdesk expense (or more helpdesk productivity, which is almost the same), as well as not having to replace lost or destroyed tokens. Next best thing to money in the bank.

Compliance

Virtually all regulation directed towards making data protection more effective addresses access controls. The people who wrote regulations have gotten this pretty much correct: Passwords are a very weak link that is protecting some highly valuable and sensitive information.

Many regulations, including PCI, require two-factor authentication for administrative access, and in many cases even for end-user (non-administrative) access. While tokens and digital certificates qualify as two-factor, biometrics are a heavy favorite, particularly since it's far more difficult for a user to lose his or her finger or eyeball than a security token.

Given the growing popularity of biometrics for access controls, if your current regulation doesn't require biometrics today, it still might someday.

Emergency Identification

There are times when persons need to be identified, but are in no position to assist with the process. In some ways, this is a little bit like dealing with uncooperative persons, except here we're talking about people who are incapacitated.

In the case of an unconscious or incapacitated person in need of medical help, proper identification is sometimes critical to saving that person's life — and you can't always count on finding some form of identification on the victim. For example, someone struck by a car while riding a bicycle may not have ID on his or her person; for that matter, homeless people often have no ID with them at any given time.

Although biometrics has a tremendous potential to help in these situations, we currently have no national database that could be used for such a purpose. There are discussions about how to generate health-care records that would include such information strictly for use in health care — but serious privacy questions will need to be answered before this problem can really be solved.

No Identity Theft

This benefit is pretty obvious, given the nature of biometric authentication and identification, but since identity theft is the fastest growing crime in America — possibly the world — we thought it was worth separate mention here.

Identity theft depends on a thief being able to use enough private information about you to impersonate you and use your credit, as well as other assets. The typical approach is to apply for credit using your correct name, Social Security number, and other information, providing *an address that the thief controls* as the billing address. The faker then receives credit cards, checks from your bank, and other financial instruments at that address.

One of the biggest faults in the system here is that the organization issuing credit or other financial documents is not sufficiently authenticating the person applying for credit in your name. Instead, someone in-house who should know better is using information that is *assumed* to be private and known only to you (such as your Social Security number and mother's maiden name). Since these items can often be found in dumpsters, online, and a host of other public places, that assumption doesn't hold nearly as well as we'd like.

If we introduce any level of biometric authentication into this process — and we are careful about the registration process — stealing identities suddenly becomes a whole lot harder. Pretty simple, huh?

In practice, credit-card companies are so focused on getting new cards into the hands of eager consumers of credit that they would never consider putting a biometric registration and authentication process in place. Doing so would seriously slow down the frantic pace of card issuance and revenue. We aren't being cynical here — honest — we're paraphrasing from actual conversations we've had with card companies. The money they make from new customers more than offsets the money they lose from identity theft, so they don't feel a need to change the system.

Coolness

Let's face it. Biometrics are just *cool*. Walking up to a door, glancing at the camera and having the door open because your iris was a registered user of the system is *James Bondian* no matter how you, uh, *look* at it. Houses that vary the room temperature based on knowing who is in the room, computers that black out portions of the screen because they recognize that the person looking over your shoulder isn't allowed to read those parts, refrigerators that allow the teenager to grab a midnight snack, but remind you of your diet and remain locked to you at that hour — all super-cool stuff.

The thing is, the stuff we've listed here isn't even scratching the surface of all the cool things we're likely to see from biometric identification and authentication systems. These technologies have only been in use for a short time and have yet to see really mainstream applications, so we're just using our limited imaginations to describe a world where computers, refrigerators, children's toys, and ATMs all know who we are when we walk up and consider using them.

Sure, there are privacy issues, they aren't perfectly accurate and some aren't the most convenient to use — but that doesn't really detract much for their coolness. We're confident that all these issues will be addressed (as well as additional ones we haven't considered yet), to allow a technology with such potentially positive impact on our lives to progress — and *allow* our coffee makers to address us by name in the morning and make our coffee the way we like it.

Part V
Appendixes

The 5th Wave   By Rich Tennant

"It's the break we've been waiting for, Lieutenant. The thieves figured out how to turn on one of the stolen video conferencing monitors."

In this part . . .

In Part II, we discuss the different types of biometric technology, and here we include a full side-by-side comparison of these technologies just like you find in consumer magazines.

We've tucked in an appendix for IT professionals who need a quick read on physical security. Biometrics technology is one of those "convergence" technologies that brings together security pros in the IT and physical/facilities professions.

The world of biometrics practically has its own language — and sometimes you just have to talk the talk. Because we don't want you getting tripped up in it, we've included a listing of the key terms that you'll commonly run into.

Appendix A

Comparing Biometrics Solutions

• •

*B*oth authors were avid readers in their youth, and enjoyed poring over statistics, specifications, and product-comparison charts. We recognize and appreciate the unfamiliarity of a new technology, and how a visual and graphic comparison can help the reader easily understand the differences and similarities of various technologies — in this case, biometrics technologies.

The two-page chart in this appendix compares most of the biometrics technologies discussed in this book — in particular, their costs and other factors — all in one place.

You should use this chart to help clarify your practical understanding of how different biometrics technologies compare to each other, but not as the sole means for choosing a technology. If you're considering biometrics as a tool for better control of access to computers or physical spaces, this chart might help you easily eliminate technologies that would be a poor fit. (***Note:*** In the chart, H=High; M=Medium; L=Low.)

If you're doing some of your comparison shopping here, then we suggest that — after you've narrowed the field to a few types of biometrics technology — you turn to the chapter(s) that describe those biometrics in more detail. Those chapters are

- ✔ Chapter 4: Fingerprint, palm scan, hand-vein scan, and hand sonar.
- ✔ Chapter 5: Signatures of all types.
- ✔ Chapter 6: Iris, retina, facial, and ear.
- ✔ Chapter 7: Everything else: speech, DNA, gait, and typing. (The chapter also covers brain wave biometrics.)

Biometric Type	Cost	Moore's Factor	Maturity	Subject Uniqueness	Weaknesses
Fingerprint	$	L	H	H	Easily spoofed
Palm scan	$$	L	H	M	
Hand-vein scan	$$$	L	M	H	
Hand sonar	$$$	L	M	H	
Iris image	$$	L	H	H	Only one vendor
Retina image	$$$	L	H (1)	H	
Facial image	$	M	M	M	
Facial thermo-graph	$$	M	M	L	Not accurate by itself
Ear Imaging	$	L	H	M	Easily spoofed, hidden
Typing dynamics	$	L	H	M	Fatigue, injury
Signature image	$	L	H	M	Forgeries possible
Signature w/ 2D accel.	$	L	H	H	Need many samples
Signature w/ stylus press.	$$	L	H	H	
Signature w/ 3D accel.	$$	L	H	H	
Speaker recogni-tion	$	L	H	M	Spoofing, playback
DNA	$$$$	H	H	H	Very slow process
Gait	$$	M	L	M	High false negatives

Counter-measures	Collect-ability	Reliability	Perma-nence	Conven-ience	Acceptance
Guards or attendants	H	H	H	H	M
	H	M	H	H	H
	H	H	H	H	H
	H	H	H	H	H
	H	H	H	H	H
	M	H	H	M	M
	H	M	M	H	H
Additional credentials	H	M	M	H	H
Add'l cred.; RFID ear tags	M	M	M	M	M
Coffee and aspirin	H	L (2)	L	H	H
Guard or attendant	H	L	L	H	H
Additional credentials	H	H	L	H	H
	H	M	L	H	H
	H	H	L	H	H
Additional credentials	H	L	M	H	H
Additional credentials	L	H	H	L	L
Additional credentials	M	L	L	H	M

Column definitions

Here are the descriptions of the columns in the tables on the preceding pages:

- ✔ **Biometric type:** This is the specific approach to biometric measurement used. Many of the biometric types discussed in this book appear in the table.

- ✔ **Cost:** The relative cost for biometric hardware, software, or other means of equipment. The symbols ($, $$, $$$, $$$$) are relative only, and don't reflect a specific price range; $ means the biometric technology is inexpensive; $$$$ means it's too costly for most business purposes. (Alcohol, dessert, and tip not included.)

- ✔ **Moore's Factor:** This refers to Gordon Moore, the originator of Moore's Law that cites the exponential increase in the power of microprocessors. Here, we mean the relative amount of computing power required to support a given biometric approach.

- ✔ **Maturity:** This column refers to how well developed the technology is that supports the measurement and processing of this type of biometric. *High* would indicate high maturity; *Low* would indicate an emerging or young technology.

- ✔ **Subject uniqueness:** Here, we're talking about the probability that a biometric system will be able to distinguish any two subjects in the world. *High* would indicate solid distinction; *Low* would mean that the biometric system would have more trouble distinguishing the subjects.

- ✔ **Weaknesses:** Any weaknesses associated with a particular type of biometric measurement.

- ✔ **Countermeasures:** Steps — such as providing extra credentials — that can be taken to overcome specific weaknesses.

- ✔ **Collectability:** How easily the biometric measurement can be taken from the subject.

- ✔ **Reliability:** How reliable is biometric technology for collecting this type of biometric. Is this type of biometric free from high FAR (false acceptance rate) and FRR (false rejection rate).

- ✔ **Permanence:** This refers to the rate at which a measured biometric will change over time. *High* means the biometric does not change over time (such as fingerprint or DNA). *Low* means the biometric changes significantly over time (signatures, for example).

- ✔ **Convenience:** How easily a subject can provide a biometric measurement.

- ✔ **Acceptance:** How willing a subject is to provide a biometric measurement.

Notes

1. Retina biometric measurement is mature but giving way to iris recognition.

2. Keystroke recognition is more accurate when sampling takes place over a longer period.

Appendix B

Controlling Physical Access

· ·

*I*n this appendix, we look at physical security using biometrics, a subject that might be unfamiliar to readers with backgrounds in information technology. Physical security requires many components that you're familiar with, such as authentication and authorization, but also some concepts that don't have any real mapping to IT, such as wall thickness, sightlines, and observability. Fear not, we make some sense of all these non-electronic terms and provide some basic understanding of how biometric systems interact with physical-security systems to control access to buildings, rooms, and campuses.

Understanding General Principles

Physical security includes those aspects of a protective system that prevent access to assets via material means such as physical intrusion. In the average home, that means a fence, doors, windows with locks, and possibly the family dog — trained to sound like a killer when someone approaches the door. In more corporate settings, it will include several points of entry into the company offices, such as a gate to get into the parking area, a security elevator that requires an access key, or a door from the lobby into the inside space, and even guards watching key points or patrolling the areas.

As with information security, physical security is concerned with the confidentiality, integrity, and availability (CIA for short) of the assets protected. However, it doesn't stop there; we have to add in some aspects that information security rarely has to address directly, such as:

- ✔ **Safety:** A breach of physical security can be a risk to the safety of people in the organization.

- ✔ **Physical theft:** Usually, when people steal information, they steal a copy, and you still have the information. When they steal your office equipment, computers, or lobby art, you no longer have access to it. (We have a list of lobby art we would *like* to see stolen just so we don't have to look at it anymore, but no such luck.)

✔ **Vandalism:** Defacing property can make it pretty unsightly and potentially embarrassing.

✔ **Physical damage:** As with theft, physical damage means you no longer have access to the asset that was damaged.

✔ **Sabotage:** This can take the form of physical damage, which is usually obvious, but people with physical access to a protected space might choose to sabotage something in a way that will be discovered when they're long gone or when you might have forgotten about the incident.

The job of protecting an organization from physical threats is, in some ways, easier than protecting information assets, but in other ways, it's harder. We compare the two in Table B-1.

Table B-1	Physical Security Challenges versus Information Security Challenges	
	Information Security	*Physical Security*
Attacks easily recognized	Very difficult at times. Not all attacks look like attacks (for example, phishing).	Usually pretty easy. Social engineering attacks are the exception here.
Tools required to make the attack	Medium to good computer skills.	A hammer or a big rock.
Exposed attack surface (places that can be attacked)	Usually network-based, sometimes not well known, frequently larger than we thought.	Pretty well defined as gates, fences, doors, windows, and the like, but creative attackers will sometimes cut through walls.
Potential attacker population	Anyone with an Internet connection, all over the world, 24 hours each day.	People with physical access to the protected facility.
Average success rate	Oddly, pretty low. Most hackers try hundreds or thousands of systems before they gain access to one, but the success depends on the skill of the hacker.	Depends on definition of "success." Most folks are successful at gaining entry if they are determined, but other measures might still foil the attack.

A quick browse of Table B-1 might make it seem that physical security is a snap, compared to information security, but we would like to highlight some items here.

The tools required to perform a physical attack are actually obtained more easily than a hammer in some cases. Consider a relatively innocent example: Mike has purchased two homes in the Seattle area that, for various reasons, had all the available keys locked inside. He was able to gain entry into both houses by using items found lying around in the yard, and without breaking any windows. The point is this: Protecting entrances to buildings from *all* forms of physical attack is nearly impossible unless you're building a bunker (and even then, it's a challenge).

If people ran around the neighborhood or the office park jiggling door handles and trying every single window to see if they happened to be open or could be opened with a good hard tug — pretty much how Internet hackers test firewalls — an enormous physical crime wave would result in a lot of barred windows and steel-reinforced doors. The reason this doesn't happen is that the risk of detection and capture when jiggling a door handle is several thousand times higher than the risk of detection when *port scanning* (looking for open ports) on a firewall. Good physical security depends on multiple layers of barriers, alarms, and reactions to the threat.

Using Barriers

Creating obstacles to entry is probably the most commonly used physical-security tool, and it's simple to understand: If you make it hard to gain physical access by putting a steel-reinforced door with a great lock between the attacker and the asset, you've protected the asset. Some subtleties that can enhance this approach, especially when combined with biometrics, are worth giving some thought.

In the physical world, it's also somewhat easier to see (literally) what avenues an intruder might use to gain entrance to your facility. For instance, if you have a big window next to that steel-reinforced door, then your intruder is going to toss a rock through the window and disregard that wonderful-and-highly-protective door.

Time is not your friend

Nearly any physical barrier can be breached. Given enough time, anyone with a good drill and some diamond-tipped bits could eventually make it through a bank-vault door. This almost never happens these days because bank-vault doors are guarded by people who would object when they saw you walking up to the door with your tools and one of those loot bags that cartoon burglars always seem to carry.

With some biometric systems, the drill and bits might not be necessary, but the system might be vulnerable to "gummy fingers" or injection attacks based on getting access to the database and to the wire between the biometric sensor and the main system. Both of these techniques typically take a lot of time to set up and test (unless you get really lucky on your first try with the "gummy finger"). Therefore, you need to make sure that any would-be attacker can't get comfortable near the door (or the sensor) and that there are good "sightlines" from the entry point to the public areas. Figure B-1 shows an example of poor sightlines and visibility of a rear entrance.

Trees or shrubs can partially obscure the entrance from a parking lot to a building. In this example, an attacker could spend some time working on this door after hours without fear of being seen from the road or anyone near the front of the building. According to our friends in various police organizations, most physical break-ins occur through entrances or windows that are hidden from casual view.

Shrubbery obscures view of protected rear entrance.

Figure B-1:
An obscured rear entrance allows attackers time to gain access.

Never count on strength alone

When an attacker doesn't have time to attack your physical security, they might get through just by using greater force than your protections are designed to repel. For example, a two-ton automobile traveling at 30 mph and chained to a door handle will produce more than enough force to shear all the hinge pins in a typical door. In the opposite direction, not too many walls are built to withstand a car traveling any faster than 25 mph, and if the assets you're protecting are worth more than a stolen 1976 Chevy Vega, the attacker sees the Vega as no more than a large, expendable key to your door.

You can bolster the strength of your barrier defenses, both as a unit (by using stronger materials) and as a system (by requiring an attacker to break down more than one strong physical defense to gain access). In Figure B-2, we show a building's main entrance with concrete and steel _bollards_ (short anchored posts reinforced to provide a barrier to ramming vehicles) and anchored picnic tables that provide an outer barrier against someone attempting to gain access to the building by smashing through the doors.

As far as a biometric system is concerned, attacks on the barriers associated with an installation would be considered a _bypass attack;_ technically, the attacker is attempting to gain access by going "around" the biometric system entirely, even if that means through the wall.

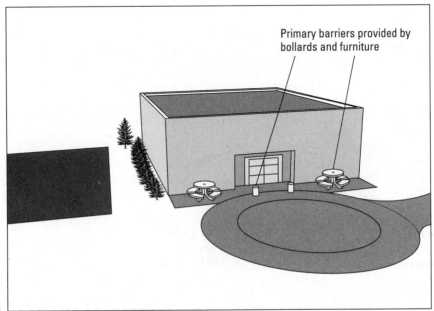

Primary barriers provided by bollards and furniture

Figure B-2:
Front
entrance,
protected by
bollards and
anchored
furniture.

It's alarming

Alarms can take on many forms, from piercing loud klaxons (which can even act as physical barriers at certain volumes) to silent electronic messages in the form of e-mail or pages that alert the people (either professional monitoring services or law enforcement) who then respond to the incident.

A fundamental difference between how information-security people work and how physical-security people work is in how they implement alarms. In an information-security capacity, we think of alarms as early warning systems, telling us that our systems are under attack. Part of the rationale here is that once someone has successfully attacked a computer system, the intruder becomes *harder* to detect than while performing the original attack. Depending on how much control the bad guys gain over the system, they might become *impossible* to detect after they get in.

Physical security works in almost the exact opposite way: Attackers might be almost impossible to detect while they're testing the system (jiggling door handles, for instance, might look just like a legitimate person trying to get in), but once they breach the physical defenses, they're actually easier to identify. Because physical attacks usually involve some level of destruction, force, or general mayhem to the facility, detecting a break-in of this type is as simple as inspecting the damage. Damage to an information system might be subtler — so much so that it's nearly impossible to detect.

Going back to the obscured doorway example in Figure B-1, an attacker using a fake finger with a stolen fingerprint on it might need to try several times to get the fake finger to work. The information-security systems should start sending alarms at some point during this process, telling people that an intruder is likely trying to get in through that door. If the attacker gets fed up with trying to get the fake finger to work and uses a big SUV to yank the door open, the physical-security systems should start the audible alarms and electronic messages to tell people that this door was opened without first being unlocked.

Preventive measures

The subtlest preventive measure is (oddly enough) just having good physical security. Many of the aspects of good physical security are visible to passersby and show that the installation has taken some significant level of care in protecting the assets. In some cases, this will include visible video surveillance (though that should always be combined with not-so-visible video surveillance) and alarm systems. For the slow-on-the-uptake criminals (and there are a *lot* of these), you might even have signs posted that describe that the facilities are monitored, patrolled, or even (simply) protected by an installed alarm system.

Reacting to attacks

The final line of defense for protecting your assets is by no means the last one you want the attacker to encounter, is the reaction to their physical attack. A reaction or response to a physical attack serves three purposes:

- ✔ **Cost:** To make the cost of attacking the resource greater to the attacker than the potential rewards.

- ✔ **Prevention:** To attempt to foil the attacker's attempt to damage or steal the asset by convincing them that the reaction will be swift and unpleasant.

- ✔ **Control:** To regain control of the physical asset as quickly as possible to prevent further loss of or damage to the asset.

The reason we say that this should not be the last defense the attacker encounters, is that the _threat_ of a response should be made clear to any potential attacker before anyone actually carries out an attack. The idea is to stop attackers before the attack gets far enough to provoke a physical response—if possible, to deflect them in the first place.

Excessive response

If you're going to put a threat in place, be careful not to over-promise or under-deliver your various responses to incidents. For example, if your sign says you have video surveillance and that you will detain trespassers, expect a potential attacker to test what it says. A few trips across the lawn without incident will put the attacker at ease about getting in; _being detained_ will have the opposite effect.

On the other hand, it's also possible to over-respond. In the information-security realm, specifically with intrusion-detection systems, we're used to multiple false alarms per day (sometimes per hour); we just take them in stride, because we know that's how these systems work. If your biometric system starts alarming every time someone fails to authenticate twice, and the guards are hustling back to the rear entrance several times each day, it's not going to sit well with the guards (or the poor sap who got a paper cut this morning and can't get a good read on his fingerprint). The subsequent (required-by-policy) manual authentication by visual verification of the user, usually by a manager, won't win any friends either.

Combining Efforts

For biometric systems that control access to physical assets, you'll be working closely with the people who traditionally have been in charge of your facility's physical security. Our advice is to listen carefully to what they tell you about how things work. You're integrating new technology with ideas that people have been using since the invention of doors, fences, and gates; the people who study the physical side of things have a field with a couple more thousand years of maturity than the high-tech stuff.

In our personal experience, all that maturity in the field of physical security also means that if we pay close attention, we might learn something from the physical-security folks that helps us do our information-security jobs better. Keep in mind that information-security people didn't invent the concept of "defense in depth" — we just took a few years to rediscover it and apply it to information rather than to literal moats and castles.

Appendix C

Glossary

AAA: Authentication, authorization, and accounting. *See also* authentication *and* authorization.

AAAA: Authentication, authorization, accounting, and audit. *See also* audit records.

acceptability: A characteristic of biometrics that refers to the willingness of subjects to use a biometric system.

access control: The practice of controlling which subjects are permitted to access which objects.

accounting: The process of tracking the use of system resources. *See also* AAA.

accuracy: A characteristic of biometrics that refers to how well a biometric system can distinguish between subjects.

administrative controls: Controls such as policies and procedures that define permitted and forbidden behaviors and events.

asset: A tangible or intangible object of value that is owned by a person or organization.

attack: The act of carrying out a threat with the intention of harming an asset.

audit records: Recordkeeping entries that provide an independent record of changes and queries that were performed against a dataset.

authentication: The process of making an assertion of identity that includes some proof of identity.

authorization: The process of approving access to an asset, based on verification of one's identity.

availability: The characteristic of an asset that is ready for use.

biometric passport: An international passport that contains a small RFID memory chip that contains biometric and other information related to the subject.

bypass attack: An attack in which an intruder attempts to circumvent controls so as to access a system.

change management: A formal business process in which all technical changes in an environment are formally reviewed before being implemented.

CIA: Confidentiality, integrity, and availability. Shorthand for the things that information security professionals must ensure in systems they are responsible for. *See also* availability, confidentiality, integrity.

circumvention: The process of bypassing a biometric system by some means and acquiring access to the protected systems or assets.

collectability: The relative ease with which biometric measurements can be made.

Computer Matching and Privacy Act of 1988: A U.S. law that defines how the U.S. federal government can collect and use information about U.S. citizens. This law is an amendment of the Privacy Act of 1974.

confidentiality: The characteristic of an asset that is protected from unauthorized access and harm.

configuration management: A formal business process in which all technical changes in an environment are recorded.

Data Protection Directive: *See* Directive 95/46/EC.

decryption: The process of transforming encrypted data back into its original form.

defense in depth: A practice in which more than one control is used to protect an asset.

deoxyribonucleic acid (DNA): The uniquely patterned genetic substance found in the living cells of all living organisms.

detective controls: Controls that detect an activity or event.

deterrent controls: Controls that discourage a subject from attempting unwanted activities.

digital signature: A cryptographic method used to ensure the integrity of a message or document.

Directive 95/46/EC: A law in the European Union Constitution that defines standards on the collection, storage, and dissemination of citizens' personal information.

DNA biometrics: A biometric technique that employs the collection and comparison of a subject's DNA. *See also* Deoxyribonucleic acid.

ear biometrics: A biometric technique that primarily employs the measurement of a subject's ear.

Elastic Bunch Graph Matching (EBGM): A recognition method used in facial biometrics that identifies landmark features of a face, such as the corners, top, bottom, and the center of the eyes.

Electronic Protected Health Information (EPHI): Health-related information about a particular person, regulated by HIPAA. *See also* Health Insurance Portability and Accountability Act.

electronic signature: A digital representation of a subject's identification that is affixed to a document or transaction.

encryption: The process of transforming data so it cannot be read by anyone who does not possess a decryption key.

enrollment: The process of initial registration. In a biometric system, this would consist of the subject providing several biometric samples that are paired with the subject's proven identity.

enrollment fraud: An attack in which an intruder attempts to enroll in place of a real, authorized subject.

equal error rate (EER): The point at which the FAR and FRR are equal. *See also* false acceptance rate and false rejection rate.

Executive Order 12333: An executive order signed by U.S. President Reagan in 1981 that provides clarification on laws regarding the collection of personal data by U.S. intelligence agencies.

facial biometrics: A biometric technique that employs the measurement of a subject's facial geometry.

facial thermograph: A biometric technique that employs accurate measurement and graphic representation of the heat given off by a subject's face.

fail closed: A failure of a control in which all events (authorized and unauthorized) are denied.

fail open: A failure of a control in which all events (authorized and unauthorized) are permitted.

Failure to Acquire (FTA): A failure of a biometric system to acquire a biometric sample from a subject.

Failure to Enroll (FTE): A failure of a biometric system to enroll a subject, usually due to a wide variance in collected samples.

False Acceptance Rate (FAR): The percentage of unauthorized users who are incorrectly granted access.

False Rejection Rate (FRR): The percentage of authorized users who are incorrectly denied access.

fingerprint: The unique pattern of friction ridges found on a human finger.

fingerprint scanner: A biometric device used to obtain a subject's fingerprint.

gait: The manner in which a subject walks, including the motion of arms, legs, and torso while walking.

gait biometrics: A biometric technique that employs the measurement of a subject's gait.

hand-vein scanner: A biometric device used to obtain the pattern of veins from within a subject's hand, usually by shining an infrared light on it.

hash: A cryptographic function that provides a fixed-length output from a variable-length input. Unlike encryption, a hash cannot be reversed to obtain the original input data.

Health Insurance Portability and Accountability Act (HIPAA): A U.S. law that defines standards of protection for Electronic Protected Health Information. *See also* Electronic Protected Health Information.

Homeland Security Presidential Directive 12 (HSPD-12): A U.S. presidential directive that requires common identification standard for all U.S. federal employees and contractors; the standard uses two fingerprints and a photograph as the samples for each individual.

identification: The process of making an assertion of identity that does not require additional assertions. The equivalent of knowing who someone is by looking at a photograph without any label.

informational privacy: The relationship among the collection, use, laws, and users' wishes regarding the collection and use of personal information.

integrity: The characteristic of an asset that is protected from unauthorized changes.

iris: The round membrane of the eye that controls the size of the pupil, and which forms the colored portion of the eye.

iris scanner: A biometric device used to obtain the image of a subject's iris.

Linear Discriminant Analysis (LDA): A recognition method used in facial biometrics to identify distinguishing characteristics between subjects.

multi-factor authentication: A method of authentication in which two or more items are provided as proof of a subject's identity.

multimodal biometrics: A method that uses more than one type or kind of biometrics to establish greater accuracy for authentication or identification purposes by providing additional information to the authentication or identification process.

palm scanner: A biometric device used to obtain an image of the prints and creases of a subject's hand.

performance: In biometrics, the electric and computing processing power required to measure and compare biometric measurements.

permanence: The relative degree and rate of change that happens to the biometrics of subjects being measured.

Personally Identifiable Information (PII): Any piece (or pieces) of information that can be used to uniquely identify a person.

preventive controls: Controls that prevent unwanted events.

Principal Components Analysis (PCA): A recognition method used in facial biometrics that compares the relative distances between landmark features of a subject's face.

privacy: _See_ informational privacy.

Privacy Act of 1974: *See* Computer Matching and Privacy Act of 1988.

proof of life: A capability of biometric devices that attempt to detect whether the source of the biometric being measured is living.

pupil: The small opening in the eye through which light passes to reach the retina.

REAL ID Act of 2005: A U.S. federal law that requires standard issuance procedures for U.S. state drivers' licenses and ID cards.

replay attack: An attack in which an intruder tries to perform an authentication by replaying data captured from an earlier observed login.

requirements: Stated characteristics of a desired solution.

retina: The layer of tissue at the back of the eye where incoming light is converted into neural signals.

retina scanner: A biometric device used to obtain the image of a subject's retina.

signature biometrics: A biometric technique that employs the collection of a subject's handwritten signature.

signature dynamics: A biometric technique that employs the collection of stylus motion used to create a handwritten signature.

social engineering: An attack in which an intruder enlists the unwitting assistance of others to obtain information that can lead to unauthorized access.

sonar biometrics: A biometric technique that uses sonar technology to collect an image of a subject's finger or hand.

speaker recognition: A biometric technique that collects a sample of a subject's speech, which can then be compared to new, unauthenticated samples.

stylus: A hand-held object resembling a pen or pencil, often used in signature biometrics.

threat: A potential activity that would, if it occurred, harm a system.

training: An essential part of a biometrics implementation, to ensure that users understand how to operate the biometric system.

Transportation Worker Identification Credential (TWIC): A biometric identity card issued to about 1.5 million workers — including longshoremen, truck drivers, merchant mariners, and port employees in the U.S.

two-factor authentication: *See* multi-factor authentication.

typing biometrics: A biometric technique that measures a subject's typing, especially the timing between keystrokes.

ultrasound biometrics: A biometric technique that collects an image of a subject's finger or hand using ultrasound imaging technology.

uniqueness: How well a particular biometric distinguishes one person from another.

United States Visitor and Immigrant Status Indicator Technology (US-VISIT): A U.S. State Department program in which fingerprints are collected from subjects at the point of entry into the United States.

universality: In biometrics, whether every person likely to be measured has the characteristic being measured.

voice recognition: A biometric technique that compares a spoken response to a sample in order to determine the identity of a subject.

vulnerability: A weakness in a system that may permit an attacker to compromise it.

Index

• C •

• K •

• L •

BUSINESS, CAREERS & PERSONAL FINANCE

Fundraising For Dummies
0-7645-9847-3

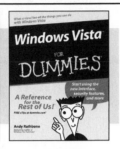

Investing For Dummies
0-7645-2431-3

Also available:
- Business Plans Kit For Dummies
  0-7645-9794-9
- Economics For Dummies
  0-7645-5726-2
- Grant Writing For Dummies
  0-7645-8416-2
- Home Buying For Dummies
  0-7645-5331-3
- Managing For Dummies
  0-7645-1771-6
- Marketing For Dummies
  0-7645-5600-2

- Personal Finance For Dummies
  0-7645-2590-5*
- Resumes For Dummies
  0-7645-5471-9 †
- Selling For Dummies
  0-7645-5363-1
- Six Sigma For Dummies
  0-7645-6798-5
- Small Business Kit For Dummies
  0-7645-5984-2
- Starting an eBay Business For Dummies
  0-7645-6924-4
- Your Dream Career For Dummies
  0-7645-9795-7

HOME & BUSINESS COMPUTER BASICS

Laptops For Dummies
0-470-05432-8

Windows Vista For Dummies
0-471-75421-8

Also available:
- Cleaning Windows Vista For Dummies
  0-471-78293-9
- Excel 2007 For Dummies
  0-470-03737-7
- Mac OS X Tiger For Dummies
  0-7645-7675-5
- MacBook For Dummies
  0-470-04859-X
- Macs For Dummies
  0-470-04849-2
- Office 2007 For Dummies
  0-470-00923-3

- Outlook 2007 For Dummies
  0-470-03830-6
- PCs For Dummies
  0-7645-8958-X
- Salesforce.com For Dummies
  0-470-04893-X
- Upgrading & Fixing Laptops For Dummies
  0-7645-8959-8
- Word 2007 For Dummies
  0-470-03658-3
- Quicken 2007 For Dummies
  0-470-04600-7

FOOD, HOME, GARDEN, HOBBIES, MUSIC & PETS

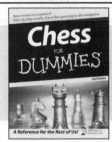

Chess For Dummies
0-7645-8404-9

Guitar For Dummies
0-7645-9904-6

Also available:
- Candy Making For Dummies
  0-7645-9734-5
- Card Games For Dummies
  0-7645-9910-0
- Crocheting For Dummies
  0-7645-4151-X
- Dog Training For Dummies
  0-7645-8418-9
- Healthy Carb Cookbook For Dummies
  0-7645-8476-6
- Home Maintenance For Dummies
  0-7645-5215-5

- Horses For Dummies
  0-7645-9797-3
- Jewelry Making & Beading For Dummies
  0-7645-2571-9
- Orchids For Dummies
  0-7645-6759-4
- Puppies For Dummies
  0-7645-5255-4
- Rock Guitar For Dummies
  0-7645-5356-9
- Sewing For Dummies
  0-7645-6847-7
- Singing For Dummies
  0-7645-2475-5

INTERNET & DIGITAL MEDIA

eBay For Dummies
0-470-04529-9

iPod & iTunes For Dummies
0-470-04894-8

Also available:
- Blogging For Dummies
  0-471-77084-1
- Digital Photography For Dummies
  0-7645-9802-3
- Digital Photography All-in-One Desk Reference For Dummies
  0-470-03743-1
- Digital SLR Cameras and Photography For Dummies
  0-7645-9803-1
- eBay Business All-in-One Desk Reference For Dummies
  0-7645-8438-3
- HDTV For Dummies
  0-470-09673-X

- Home Entertainment PCs For Dummies
  0-470-05523-5
- MySpace For Dummies
  0-470-09529-6
- Search Engine Optimization For Dummies
  0-471-97998-8
- Skype For Dummies
  0-470-04891-3
- The Internet For Dummies
  0-7645-8996-2
- Wiring Your Digital Home For Dummies
  0-471-91830-X

* Separate Canadian edition also available
† Separate U.K. edition also available

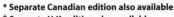

Available wherever books are sold. For more information or to order direct: U.S. customers visit www.dummies.com or call 1-877-762-2974.
U.K. customers visit www.wileyeurope.com or call 0800 243407. Canadian customers visit www.wiley.ca or call 1-800-567-4797.

SPORTS, FITNESS, PARENTING, RELIGION & SPIRITUALITY

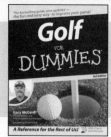

0-471-76871-5

0-7645-7841-3

Also available:
- Catholicism For Dummies
  0-7645-5391-7
- Exercise Balls For Dummies
  0-7645-5623-1
- Fitness For Dummies
  0-7645-7851-0
- Football For Dummies
  0-7645-3936-1
- Judaism For Dummies
  0-7645-5299-6
- Potty Training For Dummies
  0-7645-5417-4
- Buddhism For Dummies
  0-7645-5359-3

- Pregnancy For Dummies
  0-7645-4483-7 †
- Ten Minute Tone-Ups For Dummies
  0-7645-7207-5
- NASCAR For Dummies
  0-7645-7681-X
- Religion For Dummies
  0-7645-5264-3
- Soccer For Dummies
  0-7645-5229-5
- Women in the Bible For Dummies
  0-7645-8475-8

TRAVEL

0-7645-7749-2

0-7645-6945-7

Also available:
- Alaska For Dummies
  0-7645-7746-8
- Cruise Vacations For Dummies
  0-7645-6941-4
- England For Dummies
  0-7645-4276-1
- Europe For Dummies
  0-7645-7529-5
- Germany For Dummies
  0-7645-7823-5
- Hawaii For Dummies
  0-7645-7402-7

- Italy For Dummies
  0-7645-7386-1
- Las Vegas For Dummies
  0-7645-7382-9
- London For Dummies
  0-7645-4277-X
- Paris For Dummies
  0-7645-7630-5
- RV Vacations For Dummies
  0-7645-4442-X
- Walt Disney World & Orlando
  For Dummies
  0-7645-9660-8

GRAPHICS, DESIGN & WEB DEVELOPMENT

0-7645-8815-X

0-7645-9571-7

Also available:
- 3D Game Animation For Dummies
  0-7645-8789-7
- AutoCAD 2006 For Dummies
  0-7645-8925-3
- Building a Web Site For Dummies
  0-7645-7144-3
- Creating Web Pages For Dummies
  0-470-08030-2
- Creating Web Pages All-in-One Desk
  Reference For Dummies
  0-7645-4345-8
- Dreamweaver 8 For Dummies
  0-7645-9649-7

- InDesign CS2 For Dummies
  0-7645-9572-5
- Macromedia Flash 8 For Dummies
  0-7645-9691-8
- Photoshop CS2 and Digital
  Photography For Dummies
  0-7645-9580-6
- Photoshop Elements 4 For Dummies
  0-471-77483-9
- Syndicating Web Sites with RSS Feeds
  For Dummies
  0-7645-8848-6
- Yahoo! SiteBuilder For Dummies
  0-7645-9800-7

NETWORKING, SECURITY, PROGRAMMING & DATABASES

0-7645-7728-X

0-471-74940-0

Also available:
- Access 2007 For Dummies
  0-470-04612-0
- ASP.NET 2 For Dummies
  0-7645-7907-X
- C# 2005 For Dummies
  0-7645-9704-3
- Hacking For Dummies
  0-470-05235-X
- Hacking Wireless Networks
  For Dummies
  0-7645-9730-2
- Java For Dummies
  0-470-08716-1

- Microsoft SQL Server 2005 For Dummies
  0-7645-7755-7
- Networking All-in-One Desk Reference
  For Dummies
  0-7645-9939-9
- Preventing Identity Theft For Dummies
  0-7645-7336-5
- Telecom For Dummies
  0-471-77085-X
- Visual Studio 2005 All-in-One Desk
  Reference For Dummies
  0-7645-9775-2
- XML For Dummies
  0-7645-8845-1